INSIDER at SAC

Operations Analysis Under General LeMay

STRATEGIC AIR COMMAND

World Wide Operations Analysis Conference, 1959. Left to right: Carroll Bryant, Chief, OA; Rod Smith, OA, SAC; Carroll Zimmerman, Chief, OA, USAF; Bill Blaylock, Chief, OA, ADC; Roy Brothers, Retired Chief, OA, USAF.

INSIDER at SAC

Operations Analysis Under General LeMay

by Carroll L. Zimmerman

Sunflower University Press®
1531 Yuma (P. O. Box 1009), Manhattan, KS 66502-4228

ISBN 0-89745-097-3

Technical Editor
Abigail T. Siddall

Layout
Lori L. Daniel

Contents

Preface

In September 1980 General Curtis E. LeMay (Retired), former Chief of Staff, United States Air Force, and Commander-in-Chief, Strategic Air Command, wrote a letter to me:

Dear Zim:

I urge you to write your memoirs, particularly of your time at SAC . . . it is a must. Some of the Air Force history that has been written by people who weren't there is somewhat disappointing. The story of how your team was integrated into the staff, and the results they obtained, should be preserved for future use.

I remember when we started out I suggested you urge your people to write papers, attend scientific seminars, etc., to enhance their reputations so they could go on to better jobs. As I remember no one left unduly — they were not only having fun but really accomplishing something. The story should be written. . . . I hope you will get on with the job.

Sincerely,

Curt

* * *

In May 1943 I went to Headquarters, VIII Fighter Command, North London, England, as a founding member of the Operations Analysis Office. Later I assisted in founding the following Operations Analysis offices:

IX Tactical Air Command in England and across France, Belgium, and Germany. I became the first and only Chief, Operations Analysis.

Continental Air Forces at Bolling Field, Washington, D.C., in June 1945 after Germany surrendered. I became the first and only Chief, Operations Analysis.

Headquarters, Strategic Air Command (SAC), at Bolling Field, Andrews Air Force Base, and Offutt Air Force Base, Omaha, Nebraska.

I became the first Chief, Operations Analysis, SAC, on the day SAC was founded at Bolling Field on 16 March 1946 and served in that capacity for eight and one-half years; then I was Chief Scientist, SAC for five more years.

In 1959 I was appointed Chief, Operations Analysis, USAF, in the Pentagon, and I served two tours in that capacity with a hiatus of eight years with the aerospace industry on the West Coast. For over half of my 25 years with the United States military air forces I reported directly to General LeMay.

The account in the following pages is not technical. I have attempted to make at least part of it interesting to the layman who had no direct association with the military, as well as to bring back memories for some who did. I emphasize human relationships and traits. Personal glimpses of General LeMay are inevitable: any attempt to write about that era would be incomplete without them.

Though the contribution of Operations Analysis to early Strategic Air Command is the primary topic of the book, I first briefly described my own activities from the beginning of World War II to March 1946, when SAC was established. Not described are air battles, results, attrition, damage, casualties, and so on. Instead I have recounted some of the activities, normally not found in histories, that happen in the planning for, and in the execution of, battle.

Acknowledgments

Many people have assisted me with this project, and I owe them thanks. General Curtis E. LeMay was responsible, with Ray Hemann (then with the Lockheed Corporation), for initiating the effort. Colleagues who contributed substantial information and advice, as well as written accounts, are Stanley J. Lawwill, Thomas W. Chappelle, Bruce Stafford, Richard J. Camp, Edgar Berdahl, Edward Winger, Frederick F. Fennema, and Glen Hadsell.

Other individuals, who are listed in Appendix B, assisted as well, providing facts, correcting or corroborating my own memories of events, or recounting their own experiences while working with General LeMay, I am grateful to them all.

Stan Lawwill and Ray Hemann provided valuable assistance in reviewing and editing the manuscript, and my wife, Modie, gave many hours of counseling, editing and typing. I could not have succeeded without them.

Many thanks to Mrs. Abigail Siddall for her dedicated and excellent editing.

Chapter 1

Pearl Harbor to May 1943

Whitman College

I was head of the Physics Department at Whitman College, Walla Walla, Washington, when Pearl Harbor was attacked. At about 10 a.m., Pacific Coast time, on Sunday, 7 December 1941, I was walking our Scotty, Kiltee, on the Whitman Campus. Bill Hart, a French professor, saw me and told me of the event. I went home and listened to the radio. Modie, my wife, did not learn of it until she arrived home from church. Besides the overall implications, there was much concern at Whitman, because we had so many students from Honolulu. Communication with the islands was impossible. We had blackouts for quite a few nights until the military situation became better defined. I'm certain that none of us really understood the implications and the consequences.

One of the first from Walla Walla to go into the military service, within my purview, was a physician we knew who received a commission of captain. His young son cried because he so wanted his father to be a sergeant. I, too, was most naive about the military and about war. I didn't know a corporal from a colonel, and I wanted nothing to do with the combat military.

I was surprised, if not dismayed, to tune in a Canadian radio station after Canada became involved. They were playing dance music and Canadians were dancing. Earlier I had been in Canada and had seen long lines of men waiting to sign up voluntarily with the military, and I had been much impressed. Now I could not understand how people could dance while some of their countrymen were engaged in combat with the enemy. Naive, eh?

I was director for Whitman for ground training in a Civil Air Patrol pilot-training program being conducted by the manager of the Walla Walla airport, who provided the flight training. I taught structures, maintenance, and meteorology. Not long after Pearl Harbor it became a basic flight-training program for the Navy. On his own initiative the Whitman College president went to the local draft board and received assurances that because of my teaching physics and my participation in the Navy flight-training program I would not be drafted.

But I became restless. Several inquiries about doing more direct war work came, and offers were made. One was from the Great Lakes Naval Training Center. Another was to do research in radar at the Massachusetts Institute of Technology; that interested me very much. The Whitman president agreed to a leave of absence "for the duration" if I would write in my application that not only would I want to return to Whitman, but I would expect to.

On V-J Day I was in San Francisco. On the first available commercial aircraft I went to Walla Walla. I wanted nothing more than to return to Whitman College. The president who had granted me leave of absence had retired, and the new president was from outside. Dr. Frank Haigh, the Dean of the College of Liberal Arts and thus my boss, said that he was pleased that I wanted to come back. But he gave me some advice: I should insist on a full professorship, and I should get it in writing.

I talked with the new president and asked for a full professorship, assuring him that salary was a minor consideration, but rank a major one. After a while I got a letter from him saying he could not do it. Though disappointed, I held no prejudice; the decision was his to make. But I really burned when I got a copy of the new president's letter to the alumni. It said that he was sorry to lose Mr. Zimmerman, but they could not compete with government salaries.

Some months after my meeting with him, the president did probably the most considerate thing he could do for Whitman College. He died.

Massachusetts Institute of Technology

I went to MIT in May of 1942 and was assigned to the Airborne Division at East Boston Airport to work on the gun-laying radar for the tail of a B-29.

Col. Stud Wright was assigned to the Airborne Division to fly an aircraft as a target and to do radar research. Stud was most competent. He had an idiosyncrasy of inserting curse words in the middle of multisyllable words. He always said, "The Japan-goddam-ese." His aircraft had the following letters painted on its nose as a name: MIGDT. That meant the Massachusetts Institute of God Damn Technology. He remained a close friend after his retirement.

General Curtis E. LeMay, Hq SAC, 1950s.

I was a very early user of mathematics modeling and fancy computers for simulation of military situations. In the beginning there were problems with the users of my results. They could be divided into two main categories, those who thought the results of studies were useless if computers were involved, and those who thought that computers could do anything, including telling what the combat crews had for breakfast. To help both groups understand better the value of simulation, I often told stories like the following: A simple surveillance radar built at MIT was installed on a ship. The first time the admiral was given a demonstration, it worked. The operator pointed out that at a given azimuth and distance there was a ship. He said it was a destroyer, which he estimated from blip size. The admiral asked, "What's its number?"

The project next to mine at East Boston Airport was GCL, Ground Controlled Landing. After several aircraft were directed into the ground, they changed the name to GCA, Ground Controlled Approach.

We all know the tremendous value demonstrated by radar surveillance and GCA, when the equipment was not required to do everything. Math modeling and simulation were of great help, if one understood their capabilities and shortcomings. The acronym GIGO for computers is well known — Garbage In, Garbage Out. A friend suggested that, to some, GIGO had another meaning — Garbage In, Gospel Out.

Chapter 2

European Theater of Operations May 1943 to June 1945

In England

I was enlisted into the mysteries of Operations Analysis in May 1943 when I went to England to VIII Fighter Command headquarters as a member of the second U.S. Army Air Corps Operational Research Office. The first office had been established at VIII Bomber Command at High Wycombe in England in September 1942. (That headquarters took over a fashionable girls' school, though only after Winston Churchill personally intervened. The Americans slept in the girls' dormitory, where above each bed was a sign, "IF MISTRESS IS DESIRED, PRESS BUTTON." The Americans wore out the system the first night.)

What is Operations Analysis? And what is the difference between Operational Research and Operations Analysis, and between Operations Analysis and Systems Analysis? I shall permit some light to escape from under that bushel basket. There is no difference between Operational Research and Operations Analysis. Once Alain Enthoven wrote 19 pages on the difference between Operations Analysis and Systems Analysis. This proves there is not much difference, because if there were he could have said it in one paragraph.

John Harlan, Chief of the VIII Bomber Command Operational Research Office, met our group in London upon arrival from the U.S. and escorted us to the Hq VIII Fighter Command at Watford, North London. It was a British holiday and all the U.S. brass were reviewing a parade. The Materiel colonel was holding down the front office. John Harlan introduced us as the newly arrived operations analysts. Yes, the colonel thought he had heard something about that. What did we do? It was fortunate that John Harlan was with us, as none of us had the faintest idea what we did. John said something like we worked on special problems of the command, to which the dear colonel replied that they had no problems. That was my introduction to Operations Analysis.

After World War II John Harlan became a member of the United States Supreme Court. He passed away in 1976.

The only electronics equipment in the command was the communications gear, and it was being jammed. We determined that it was unintentional jamming from the German Freyas, early warning radars, and we recommended frequencies to be avoided. At ABL-15, the American-British Laboratory, a simple filter was designed and retrofitted that eliminated the problem.

The U.S. fighters were controlled by the British Home Chain radar net. Knowing where a bogey (enemy fighter aircraft) was 3 to 5 minutes ago did not seem very important to me. After much opposition I finally convinced VIII Fighter Command that they could direct their aircraft from the information from one radar set alone, and eliminate the delays. American controllers, using American communications equipment together with information from the British radar scopes, proved quite successful. This led to the modification of the U.S. SCR-584 gun-laying equipment (at our suggestion) and the MEW (Microwave Early Warning), both of which were used for close control support for bombing across France, Belgium, and Germany.

When the Ninth Air Force was reactivated in the United Kingdom from Africa, I joined IX Tactical Air Command (IX TAC), commanded by then Brig. Gen. Pete Quesada. (Later he was promoted to temporary major general on the same day he was promoted to permanent major.) We operated in direct close support of the First U.S. Army.

I made my mark early when IX TAC moved to the permanent RAF station at Middle Wallop. (There is also an Upper Wallop and a Nether Wallop.) I was a member of the contingent that went a week early. I suppose we went to ensure that the new operation would get office space and some place to sleep. The RAF was still there, so for a week we ate RAF rations (but we did not honor their working hours: we thought we should be at work before 9 a.m., and we did not think we had time for an hour off for tea each morning and each afternoon).

One picked up one's food, cafeteria-style, in a room separate from the dining room. There were two doors from the dining room to the food, one at each end of the serving tables. The British went in the far door from where they entered the dining room from the hall. When they came out the near door with their trays, they had to fight their way through the queue, which was always there. I suggested to our Hq commandant that when we took over we should go in the near door and out the far door, and eliminate that problem. That made my career as an operations analyst.

* * *

At Middle Wallop two or three of us addressed the following question: if we were the Supreme Allied Commander, where would we invade the continent? We were certain that the Germans thought it would be at Calais, so we discarded that. After much discussion we selected a place called Normandy.

Nick Smith and Leon Alaoglu made a railroad interdiction study. Laying out the railway network, we determined the minimum number of cuts required to paralyze the system, and we computed the number of aircraft required for a high probability to effect each cut, the types of bombs and fuzes, and the policing required to keep the cuts open. I marked the report Confidential and took it to Pete Quesada, who was delighted. This was exactly where the invasion was planned. The report was marked Top Secret and I was given the special Bigot clearance. This meant that I was cleared for all the plans for the invasion, except for the actual timing. Pete practically took me by the hand, like a kid with a wonderful new toy, and took me to Hq Allied Expeditionary Air Forces (AEAF) at Norfolk House, where the study was briefed.

This was the railway interdiction operation used in OVERLORD. After the war I saw the memoirs of a German army general. He credited this as one of the major factors in the Allies' success in securing, holding, and extending the beachhead.

* * *

There was to be a tremendous number of radar and communications equipments in the narrow confines of the beachhead — for searchlight and gun control, for warning, and for the command and control of the air effort. Mutual interference was inevitable. The same equipments in the same relative positions they would occupy on the beachhead were sited in the Tyne-Tees area of northern England, and I was given the responsibility for assessing the mutual interference problem. Roland Larson and Harold Balmer were put in charge of the project, which was conducted in April 1944.

We recommended some changes in deployment, and the restriction or elimination of some frequencies to be used. Recommendations were made to AEAF Hq and were adopted.

* * *

Dr. Edward Bowles visited my group at Middle Wallop. He was senior scientific advisor to Hap Arnold, and he had considerable wallop. (Forgive me!) He had the authority to recommend dismissal or reassignment of ranks through major general.

We spent hours in session. I told him I categorized signal officers into two groups, those who know only wire and pigeons and were suspicious and afraid of anyone who knew more so did all they could to thwart contributions by such individuals, and those who knew only wire and pigeons but realized it and welcomed any assistance they could get.

We recommended the modification of the SCR-584 gun-laying radar equipment to enable direct control of aircraft for bombing, as well as surveillance and control. This was done, and the modified SCR-584 was early on the Continent. Because of the lack of experience of direct control and the reluctance of pilots to receive direction from the ground, the first results were disappointing. One of my men, Roland Larson, was then assigned to one of the first units. He did an outstanding job and the SCR-584 became an important tool.

The success of this program led to the design of the MEW equipment. It was so named to camouflage its true purpose. (The large radar research effort at MIT was named the Radiation Laboratory for the same reason.) We helped train and get operational the first MEW in Europe, and one of my men, Capt. (later Maj.) Edwin A. Andress, was named commander of the first unit on the Continent, consisting of some 300 men. He was awarded the Legion of Merit just prior to D-Day.

* * *

My group was given the responsibility for assessing the reconnaissance capability of the Ninth Air Force prior to D-Day. Dr. Harold Wyckoff was in charge. He did an excellent job, but I changed one of his early draft recommendations: in discussng organization he said something like, "Needed in this position is an officer of real ability and sufficient rank. Perhaps two men are needed."

* * *

Most of the radar equipments used in the mobile ground air warning and control system are very much dependent on site for efficient operation. Nearby hills should screen objectionable permanent echoes, but low cover suffers as a result of any screening.

In December 1943 AEAF headquarters charged us with the responsibility for selection of radar sites in France in what was to be the American area of operation. Topographical and mosaic maps, made from reconnaissance photographs, were studied. Sites were catalogued and labeled as to technical characteristics and accessibility. Ratings for flatness and degrees of clutter were consolidated into the final rating given each site. Included also were short statements on accessibility, cover, permanent echoes, and screening at each site.

The study was divided into two parts. The first was the bridgehead and the territory expected to be held on D + 17. The second covered the rest of the American area up to D + 60. We trained the siting officers for our command, IX TAC, and for XIX TAC and IX Air Defense Command. We assisted them in selecting practice sites in the United Kingdom and briefed them on sites in France. A pamphlet, "Siting Aids," was prepared and given to each. Arnold McLean had the primary responsibility for this work.

We were on the bridgehead a few days after D-Day and within a few miles, usually two or three miles, from the front line across France, Belgium, and Germany. This enabled us to compare the actual sites, and the results therefrom, with our pre-selected sites. The comparison was quite favorable.

* * *

During the time our radar units were deployed in the field in England for training, I secured the services of an RAF Flight Sergeant Kay, who had had extensive experience with the British Home Chain radar as operator and supervisor. He helped with the interference tests in the Tyne-Tees area. I believe the impressions of an Englishman criticizing the work of American radar personnel are interesting and illuminating. Kay wrote:

> In view of the great hospitality and kindness shown me during my attachment to this station, any criticism of the personnel given here is offered in a spirit of construction, and not of destruction; and it is added that I learned much myself from the enthusiasm and ability to work as a team that I saw during my stay.
>
> It is obvious that the technical and theoretical training given to the men in the USA and at RAF stations over here has been impressively thorough. Preparation for their job has been patently complete, and the result is that the men have an innate feeling that they are right on the top of their job. I should imagine that this "superiority" feeling would materially enhance the quality of results during actual operations.

He went on to discuss details of operations, training, situations, and equipments.

* * *

It was difficult to help some people. I was asked to help straighten out the interference between the communications equipments on the headquarters ship to be used for the D-Day invasion operations. I attempted to secure help from the British Branch Radiation Laboratory (BBRL), a group of Americans doing radar and communications research in England.

The many telephone calls to the Navy at the headquarters ship, also to the ones requesting help and to BBRL were frustrating. We were asked for assistance on 6 April, and the work had to be completed by 20 April. The normal two-hour delay in getting a call through, plus often getting cut off immediately after saying Hello, resulted in my sending a message on 18 April to Lieutenant Colonel Sladek, whom I had failed to reach by phone. The message to Sladek was to forget about possible help from us: difficulties in communications and transportation had defeated our honest efforts. The summary in my report on the attempted exercise is characteristic of attemps to help some other persons on other problems:

> Except for the first day, 6 April, when Major Bain happened to see me in the officers' club lounge, no one approached me. In every instance I contacted the other persons.
>
> When I thought I had finally made all contacts, the job had to be completed in five days. Further delays made it two or three days.
>
> If, on my own, I could make all arrangements, any help I might give would be cheerfully offered.
>
> Everyone was willing to be helped, but no one was willing to help.
>
> BBRL made attempts to cooperate. First contacted on 6 April, it was 17 April before all was arranged, so it was natural that new requirements for them should arise.
>
> Despite having much to do, my attitude was to help if possible. It proved to be impossible.
>
> The telephone in wartime England is an unsatisfactory instrument.

* * *

In the very beginning of our work overseas, certain members of the military looked upon us with grave suspicion, considering us to be a sort of special air inspector, or super gestapo. It was difficult to convince some of them that we were there to help, and not to snoop. One of the reasons was that there were so many "hit and run" scientists who continually plagued the ETO. Most of these came for only very short periods of time and had no chance to really understand the problems of the theater. Unfortunately, some of them showed a considerable lack of tact in their dealings.

Since many of them were attached at high level, they often caused real concern at the units they visited, without giving any tangible help. This made many difficulties for those of us who had to live with the military all the time. To attempt to lessen the difficulties accompanying these "free lance" civilians, Pete Quesada set up a system in IX TAC for me to coordinate the visits of such individuals and to decide who should see General Quesada. This arrangement saved time for them, as I was in a position to tell them whom to see in the command on any phase of their interest. This close contact enabled me to inform the commanding general of the status of many projects.

Having been chief of the office, I use "I" frequently. It should be realized that without the help of my men I could have accomplished little.

On the Continent

IX TAC headquarters was on the beachhead quite early. The greatest danger to us was the shrapnel falling from our own ack-ack, which was engaging the tremendous air activity overhead. The beachhead was quite small. At first we would get up from our folding canvas cots in our tent, put on our helmets, wrap our trench coats around our shoulders, and stand erect to offer the least cross-section to above. But this got old. I would lie on my cot. I had a helmet. I would put it over my face. Then I would wonder if that was where I really wanted it. I wished I had two helmets.

* * *

I was sent on assignment to the Zone of the Interior (ZI) for two weeks, for temporary duty at Wright Field, Dayton, and then the Pentagon. (At that time Wright Field and Patterson Field had not been combined as Wright-Pat Air Force Base.)

I had a Class 2 priority for overseas flights, but when I got to the U.S. I found the priority meant nothing on commercial airlines in the ZI, and early reservations were impossible to get. So I started searching for military aircraft to get from Washington to Dayton. I found that MATS (Military Air Transport Service) flew on a schedule from New York to Los Angeles. The C-47 made stops at the Middletown Air Depot in Pennsylvania and at Patterson Field, Dayton, among others, but it did not come to Washington. At Bolling Field I found a C-47 that had some maintenance work about to be finished. It would need a flight test. I persuaded the pilot to take me, at night, to the air depot in Pennsylvania.

I phoned my wife, Modie (she was staying with her mother just south of Dayton), and told her when I was scheduled to land at Patterson. It was the first she knew I was in the ZI. I told her that she would not be permitted on the field, so she should make sure that upon landing I got a message where I could find her. She did. When my aircraft stopped on the ramp, the gent who opened the door from outside asked, "Mr. Zimmerman?" They had a car waiting to take me to the gate where Modie and my Dad were waiting.

I first saw Zimmie, who was seven months old. After several days Modie and Zimmie went with me by train to Washington, but after that I didn't see Zim again until I returned home from Europe on my birthday, 2 June 1945, about a month after V-E Day. It was quite an experience to find myself the father of an 18-month old boy, but I tried hard, and he was patient, so we soon got along.

We had gone onto the Continent in field uniforms. I was fairly certain that Paris and Brussels were ahead, so I decided to get a Class A uniform. The chore of convincing the personnel in the uniform department in the PX at Patterson Field that they should sell an officer's Class A uniform to a civilian was not easy. My AGO ID card listing my assimilated rank as colonel helped and the PX officer, a commissioned officer, finally reluctantly approved the sale. We were always in officers' uniforms without rank insignia.

* * *

Arriving back in England I went to a military airbase in southern England for transport back to the Continent. I was told nothing was available until the morning, so I checked into the visiting officers' quarters. At least my Class 2 air priority was valid again. About dusk I was paged. An unexpected flight was leaving in 15 minutes for the Continent. I was aboard.

There was still only one airstrip on the Continent, though the territory under U.S. Army control had expanded. Upon landing I had the problem of finding my unit, Headquarters, IX Tactical Air Command. All I knew was that the headquarters was in tents, and its code name was "Football." I finally persuaded a GI driver of a 2½-ton truck to help me find "Football." By the time we started out it was dark, and the greatest impediments to our travel were the displaced civilian refugees on the road with a few belongings, a few carts, an occasional cow.

At one crossroads we were wondering which way to go. I saw a sign staked along the road to the right and got out with my flashlight. The sign said "VORSICHT! MINEN!" (Take care! Mines!). We did not go that way. At the crossroads in a small town an MP told us that "Football" was two miles down the road at a lane to the left through an apple orchard. We found the lane. I got out of the truck, the GI wished me good luck, and I wished him good luck. I started trudging down the lane in a Class A uniform, intense air activity overhead, my only baggage two musette bags filled with Canadian Club, with just room for some shaving gear. The MP had not said the camp was nearly two miles down the damn lane.

Suddenly out of the darkness, "Halt! Who goes there?" I said the proper thing like "Member of the Garrison" or whatever.

"What is the password?"

"I'll be damned if I know, but maybe you remember me from Middle Wallop."

"Wallop! Right! Advance and be recognized." It

just happened that "Wallop" was the password that night. I talked him out of the countersign. To his credit, he had probably decided that anyone attempting infiltration would not be in Class A uniform, and would not be going down the center of the lane using a flashlight and carrying only two musette bags filled with Canadian Club.

* * *

We lived in tents across France, including Paris. We dug our own foxholes and pitched and struck our own tents. Across Belgium we lived in buildings. Across Germany it was back to tents part of the time, but sometimes in buildings. Being always in juxtaposition with the First U.S. Army Hq, we moved every five or six days, with one or two exceptions.

At Verviers, Belgium, Operations Analysis personnel were billeted in the house that previously had been occupied by the German town commandant. Though antidemolition teams always check such structures first, the first thing we did was open all doors and drawers to find booby traps they might have missed. The house was on the highest hill in Verviers. We had to walk downhill to the headquarters building, which had been the town hall, and to the officers' mess.

There were many buzz bombs (V-1s) launched against Verviers, some V-2s, and some rocket artillery, which we dubbed V-1½s. Most people do not realize that more buzz bombs were launched against

Hq IX Tactical Air Command, Verviers, Belgium. Front row L to R: Lt. Col. Bentley, Col. Garland, Col. Johnson, Col. Fletcher, Col. Melville, Brig. Gen. Kincaid, Maj. Gen. Quesada, Col. Kelso, Col. McClennahan, Col. Mallory, Col. Cone, Col. Taylor, Lt. Col. Parkman. 2nd row L to R: Maj. Gerry, Maj. Alley, Maj. Clark, Maj. Devine, Lt. Col. Wimer, Maj. Temske, Lt. Col. Crockett, Lt. Col. Sorrells, Lt. Col. Prout, Lt. Col. Hopkins, Lt. Col. Pritchard, Lt. Col. Morgan, Lt. Col. Warren, Maj. Rayle, Maj. Bain. 3rd row L to R: Capt. Moore, Maj. Coles, Capt. —, Maj. Gambill, Lt. Col. Gordon, Maj. Moree, Maj. Mays, Maj. Carhart, Maj. Wilson, Capt. Voorhies, Capt. Rheinhert, Capt. Haaland, Maj. Fetterman, Lt. Col. Fellows. 4th row L to R: W/O. Volkenant, Capt. Schoenberger, Capt. Hughes, Maj. Kirstead, Capt. Craig, Maj. Chunn, Maj. Davis, Maj. —, Capt. Iske, Capt. Blucher, Capt. Deniston. 5th row L to R: W/O. Yandell, Capt. Grayson, Maj. White, Capt. McCabe, Capt. Coates, Maj. Uscher, Capt. Laraneta, Capt. Pinsky, Lt. Bond, Capt. Abramson. 6th row L to R: Lt. Brooke, Capt. Hudson, Lt. Brooke, Lt. Thrasher, 5th L, Capt. Van Arman, Lt. Allen, Lt. Harris, Lt. Smeltzer, OA, Lt. Kraus, 3rd from L. 7th row L to R: Lt. McAdams, W/O. —, Carroll Zimmerman, Leon Alaoglu, Capt. —, Nick Smith, Lt. Halloway. 8th row L to R: W/O. Smith, Mark Eudey, Lt. —, Lt. —, Lt. Strong, Lt. Picatti. Last row L to R: Lt. Forstenzer, S/L. Dear, W/O. Bickford, Lt. Davis, Lt. Lunsford, Lt. Polatsek, Lt. Loring.

Hq IX Tactical Air Command OA staff in Belgium, just before the Battle of the Bulge. Back row L to R: Cpl. Davidow, Roland Larson, T/Sgt. Waldo C. Boffa, Leonidas Alaoglu, Pvt. Lettier. Front row L to R: L. S. Taylor, Ninth Air Force, IX Bomber Command, Zimmerman, Nicholas Smith, Mark Eudey.

Allied troops on the Continent than against England.

The Germans were rigidly methodical in their launchings. If we had not started shaving by the time one came over in the morning, we were late. If one had not yet come over by the time we reached a certain intersection in walking down to the mess for breakfast, we were early. The buzz bombs flew at low level. In our house high on the hill, I swear some went in our attic window at one end and out the other end.

One reason we won the war was because if things didn't go according to the book, the Germans (and also the Japanese) didn't know what to do. A single American GI was a one-man army until he was incapacitated.

When entering a room, we would immediately survey the windows and other glass areas, and look for possible self-protection areas. When a buzz bomb cut off, you had a few seconds to jump to a potential protection area before the bomb hit. If you heard it cut off, it was going to hit close. But then the Germans modified the V-1s so that they would power dive into the ground without cutting off. It was quite a sporting course.

One morning when I was walking down to breakfast I saw a woman and her two children on the street killed by a buzz bomb. They were across the street a block or so ahead, and when the V-1 cut off I had time to jump behind a stone wall. They got the full impact of the shrapnel but, luckily for me, most of the fragments in my direction dug into the ground.

In war one must train oneself to be hard as steel, noncompassionate, and unemotional. I remember slipping for a bit three times. The first time was when I saw women and children killed, or dead. The second was when I saw my first dead American. I thought of wives, children, mothers, sweethearts. One must not do that. The third was when I met a GI in his thirties with three children at home, going back to the front lines at Christmas time, after recovering from his second enemy-inflicted wound.

* * *

At the start of the Battle of the Bulge we were in Verviers. We were ordered to go to Charleroi, Belgium, not directly but by way of Holland. I was fortunate to have an ambulance-type vehicle assigned to me, enclosed and with a heater, and I started driving a devious route to Charleroi.

During the Bulge many American-speaking Germans in American officer uniforms were dropped by parachute or infiltrated behind the American lines. I was stopped by MPs asking such things as, "Who is Charlie McCarthy's mouthpiece?" "Who won the World Series last year?"

One burly MP stopped me and asked, "What is the password?" I replied, "I'll be damned if I know." He grinned and waved me on. Obviously there was no password.

At Charleroi we were assigned to a very fine house. A GI was assigned to take care of the place. The only heat was a GI pot-bellied stove. Our lad got it too hot,

HQ IX TACTICAL AIR COMMAND QUARTERS IN VERVIERS

Carroll Zimmerman

Carroll Zimmerman

Mark Eudey

and it burned its way through the floor and fell to the floor below. Fortunately, the fire was seen and put out quickly by the Charleroi fire department. An MP found me at the officers' mess and told me there had been a fire at our quarters. When I got there the GI was in bed. I asked, ''What the hell are you doing there?'' He said, ''I got confused, so I went to bed.''

When we moved into Bruhl, Germany, a suburb of Cologne, my office was assigned to the house of a famous sportsman, an Allied sympathizer who had been jailed by the Germans. The house had been occupied by the Gestapo. We found a small Gestapo camera, which I still have, some putty explosives on the wall, and a hand grenade lying on top of a bookcase in a dark corner. The pin was nailed to the bookcase. I slept in the bed that the King of Holland had used various times. This did nothing for me.

* * *

On another trip across part of France I was in my vehicle alone when an MP stopped me in a town. He was attached to an Allied civil assistance unit and he asked me to take an elderly French couple back to their home in St. Lô. St. Lô had been battered to the ground before it was taken. We found the site of their home, but it was only rubble — not even a chimney standing. They stood hand in hand looking at it for awhile. Then she said without turning her head, ''Quel dommage!'' (What a pity!)

Beyond St. Lô I came to a crossroads. I got out of my vehicle with my map and was studying it and looking at the road signs, when a vehicle stopped. A gent got out and asked if he could help. His field jacket covered the rank insignia on his collar, and he had none on his helmet, but I thought he looked familiar. Together we decided which way I should go, and I thanked him. He got back in his vehicle. As the driver pulled away I saw three stars on the bumper. I realized then it was Omar Bradley.

Many years after the war, when I was going to Los Angeles from Washington, D.C., on a commercial aircraft, I recognized five-star General Omar Bradley (USA Retired) sitting in front of me. After take-off, when the seat belt sign went out, I introduced myself to him. He was most gracious, and we exchanged a few remembrances. He was going on to San Diego.

Several days later I read in the paper that he was married in San Diego.

* * *

This may sound strange, but I was glad to get back to the Continent in the combat zone after my visit to the ZI. I had been disgusted with much I saw in the U.S. — because of my strong feelings and dedication and belief that there should be equitable participation in a war. As a civilian, I had volunteered. At both Whitman and MIT I had positions that deferred me from the draft. At Dayton I had heard a woman complaining about rationing, yet her two grown sons were making more money and drinking better liquor than they ever had before.

I went with Modie and my Dad to a night baseball game in Dayton. Midway through, the loudspeaker announced that they were going to turn off all the lights and everyone should light a match "in memory of our boys overseas." I realized that for many that would be their contribution to the war effort.

On the Continent the MPs directing traffic had a curious policy. They would make sure that you would not collide with another vehicle at an intersection, but they would say nothing if there was a pitched battle a mile down the road in the direction you were heading. I guess they assumed you knew what you were doing.

We wore uniforms without rank insignia but with "U.S." on the collar on each side. It hadn't been too many years since generals wore such insignia, and some GIs remembered. We had instructions never to initiate a salute, but always to return one. This would often save a lot of explanation. But the normal way that Brigadier General Kincaid and Major General Quesada greeted me in our tent camp was to salute, and I would always return their salutes. This made the GIs wonder what kind of a guy a major general was saluting.

* * *

The Allies were quite anxious to get a Freya, a German Early Warning radar, for obvious reasons. The British and the Americans had an agreement. If one was captured intact it would be evacuated by a British ship and would go to the British, because it was too large for an aircraft. If it were in pieces and would fit in an aircraft, it would go to the Americans. Thus it appeared that the British had triumphed again in negotiation with the Americans. But they ignored American ingenuity. When Freyas did start falling to the Americans they would quickly cut them into pieces that would fit into an aircraft. They were not random cuts.

The Battle of the Falaise Pocket was the first time the Allies eliminated a complete German army. Our headquarters moved near Mortain while intense fighting was still going on. It was foggy and rainy. We pitched our tents, dug our foxholes, and tried to go to sleep. As was usually the case, we were ahead of our artillery — hence between our artillery and the front line. Usually when we moved into a new area, a priority chore was to prepare the latrines. (This was not done for comfort but for sanitation; a GI ill from lax hygienic practices was no more useful than one wounded by the enemy.) But under these circumstances, nothing was done that day.

We learned that during the night a German tank column had come within three-quarters of a mile from us. Fortunately for us, at an intersection they took a road that did not go by our camp. The tanks had to use the roads, because the hedgerows were many, high, and formidable. The ditches behind the hedgerows were also useful in the absence of latrines.

* * *

One of my colleagues, chief of the office at IX Bomber Command, sent his first report to the commander. He waited with anticipation. It came back and on the buck slip was written, "The next time you send me something like this, send a piccolo with it so I can play it." The report was replete with integral signs, summation signs, and so on. Thus I learned early to write reports in a form that could be understood by the recipient. Some analysts have not learned this yet. Some think that the object is to impress their colleagues with their sophistication. This would be fine if their colleagues were paying their salaries.

Chapter 3

Wartime Impressions: Places

England

At Watford, North London, Dr. Taylor, chief of our Operations Analysis office, and I were billeted in a private home. Mr. Panman, the owner, was a barrister with offices in London and Coventry. He worked hard at typifying the classical image of an Englishman.

Dr. Taylor eventually moved elsewhere, but I stayed with the Panmans. They had three single daughters, all graduates of Oxford University. Also staying there was a Czech refugee family. The father and mother were about my age, perhaps a bit younger. They had two children something like 8 and 10. I had a private room and private bath, but Mr. Panman made it clear that I was not there by his choice.

I found that authorities went around and counted chimney pots on each house. They subtracted one each for living room, dining room, and kitchen, and thus knew how many bedrooms there were. Taking into account the number living in the house, they determined the number of extra bedrooms available. This was confirmed by inspection.

Central heating was unknown. At the end of World War II a newspaper polled housewives as to preferred home improvements. Central heating was not among the top ten. The Panmans had a full-time gardener, a full-time cook, and a full-time maid. They had their own private tennis court and private swimming pool. They had a delightful custom of swimming in the nude.

In keeping with the professional practice my rent was in guineas. Though guineas had not been minted since 1813, this is the way the professional gets 21 shillings, instead of 20 shillings, to the pound. I learned the hard way to give Mr. Panman the exact amount of my rent, because he never had any change, and neither did I ever get any.

Doc Taylor and I were taking a walk one day not long after we arrived. We heard an English lad sing, "My mother has told me." For you youngsters, regardless of what you think of them, the published words were, "My Momma done tole me."

The last underground train from Piccadilly Circus to Watford left at 11 p.m. One night I found myself sitting beside an attractive English WAF officer. We conversed and she asked the location of my home in the U.S. When I said "Washington," she asked, "D.C. or the state where Walla Walla is?" A five-year resident of Walla Walla being asked such a question by an English girl! I found she was married to the American, a lieutenant colonel, who had the dubious distinction of being the oldest fighter pilot in the ETO. He was stationed at the same headquarters as I, and she was on her way to visit him.

* * *

My wife sent me some popcorn and Crisco. No one in the Panman household had ever seen popcorn before. The Czech kids were thrilled because they had just heard about it the previous week in school. The entire family was in the kitchen watching me pop the first pan, including Mr. Panman, Mrs. Panman and the daughters, the Czech family, the maid, and the cook. I don't know where the gardener was. A small amount in the bottom of the pan, the popping noise, and then my lifting the lid to show a full pan of popped corn. They all looked in wonderment, if not bewilderment. The cook asked, "How do you eat it? As a vegetable or a sweet?"

The Panmans had a large electric refrigerator (such refrigerators were rare), and they permitted me to keep some Grahams Lager beer in it. That was as close as I could find to my taste in beer. Most beers were weak, flat, bitter (Guinness), and warm. One of my airmen described English beer as stud urine with the foam blown off by the north end of a horse going south.

Two nights a week Mr. Panman stood fire-watch at his downtown office. On those nights I stood fire-watch at the Panmans with one of the daughters. There was some activity. When the sirens sounded we would get up and patrol a specifically assigned area.

The Panmans thought I was not very sociable when, after partaking several times, I refused high tea at 4:30 or 5 in the afternoon. That was not compatible with relatively good food at the U.S. officers' mess at six o'clock. I should hasten to say that Mrs. Panman and her daughters were dolls. She was a daughter of a Member of Parliament. The charm and warmth of those four more than made up for Mr. Panman's aloofness, artificiality, and perhaps downright hostility. I'm glad I had the opportunity to live with an English family, even though the Panmans were not at all typical of the masses.

* * *

Once a major was driving me in a jeep across the English countryside to some installation. Suddenly he pulled off the road and stopped, saying something like, "Look at that stupid farmer. He is starting to fertilize the field from that side because that is where his great-great-grandfather started." Anyone who has not experienced it cannot appreciate the influence of tradition in England. The farmer with a team of two horses had a wagon filled with powdered fertilizer. With a shovel he was throwing the fertilizer into the wind, and much of it was blowing back in his face. He and the horses and wagon were white. Had he started on the other side of the field he could have thrown downwind. This would have saved him and his horses from contamination, and he would have been better able to control the spread.

* * *

On a tour of the British Home Net radar chain, I visited the Isle of Wight off the south coast of England. It is a resort island. It was August, and we were in an open personnel carrier. I had on a sweater, a field jacket, and a trench coat and was still freezing. What really annoyed me were all the palm trees. Apparently palm trees do not particularly need warm weather, but they must never freeze. With the warm ocean currents it never got to freezing on the Isle of Wight. Ten different varieties of orchids grew on the island. One morning at one of the resort hotels where we stayed, we had kippered herring, grilled tomatoes, and baked beans for breakfast.

* * *

British vehicles were sometimes unusual. On the Continent I had befriended a British military unit, so they gave me a vehicle. It had been a mobile radar set but they had removed the radar, the lorry (truck) bed, and the cab. What I had was an engine and chassis with one seat, which could hold two besides the driver, a windscreen (windshield), and a bonnet (hood), but the rest was open air. I suspected it was surplus because it was so underpowered that it could not make it across the fields, where it was supposed to go for siting off the road. Stripped as it was, it had trouble enough on some hilly roads.

At one time on the Continent my office had three vehicles: an ambulance-type American vehicle, this British one, and a liberated German Opel. After V-E Day, when we were getting ready to start home, there was no trouble getting rid of the American vehicle and the Opel, but no one would consider taking the British vehicle. So one day I had one of my airmen drive it to an American supply depot. I followed in my own automobile. While he drove just inside the gate, I waited outside with the motor running. The airman ran to my vehicle and I took off as fast as my wheels would take me.

While in England, I had an RAF bicycle that I got on reverse lend-lease. If I lost it, it would cost me $55. If a member of the RAF lost an identical bike, it would cost him $27.50. I do not think I was supposed to know that. While on the Continent our field desks were a hunk of wood with wooden legs folding on normal hinges. I forget now the amount that each desk cost the Americans on reverse lend-lease, but it was a ridiculously high figure.

* * *

It was, and still is, difficult for me to understand British democracy. During my first few months there Parliament passed, among others, the following laws:

If a man gave his wife 20 bob (bob is slang for shilling — at that time worth 24 cents) a week for expenses, any money she had remaining at the end of the week was legally his.

Parliament refused to pass mandatory testing of cows for TB, even though the incidence of TB was probably the greatest in the world.

Food was rationed for horses, but not for racehorses.

* * *

Meat was severely rationed for the masses, but if one had five pounds (then worth $24) there were many places in London where one could get an excellent meat-course meal. Several months after arrival I was having dinner in London with an American friend who was with the National Research Council. When we went to the cashier to pay our checks, he found he was short of cash. I offered to lend him some, but he refused as he had a crown, which is worth five shillings. The cashier would not accept it. We found that superstition says that a crown brings bad luck, apparently even more so than some used to attribute to our $2 bill. The cashier said that my friend had an honest face; he should bring the balance in sometime.

Crowns are minted only in coronation years, and are not used as regular tender. The next morning I went to a bank and purchased a crown. I have had it continuously with me in waking hours ever since. I decided that anything unlucky for the British would be a lucky piece for me. I wish I had kept count of the number of Englishmen who have seen a crown only once — mine. This includes a bellhop at the London Hilton who saw it among my change on the dresser, and asked what it was.

* * *

The British public telephone required two pence (pennies) — the large cartwheels. To make a call, besides inserting the two pennies it is necessary to push button "A" and button "B" in the correct sequence. One has to push button "B" after the phone is answered. If you do not, the called party cannot hear you, but you can hear him. Forgetting to push button "B" made a dull converstation. If no one answers, pushing button "B" would get back your two pence. Once at a pay phone in the underground (subway) at Piccadilly Circus, my party did not answer. When I pushed button "B" I got three pennies back. I had inserted two. I tried again with the same result. I did this until I got tired or perhaps I thought my pockets would hold no more cartwheels. Though the payoff was small, it was the best slot machine I ever played.

* * *

Liquor for us was strictly rationed. Hence the IX TAC officers' club only sold liquor one night every two weeks. We would all donate our share and have a party every two weeks, with a band and guests. On the alternate Saturday nights the medics came through with some very good punch. It was very smooth, and most of the English girls would want more. We danced to a juke box and had some very good times at these "off" weekends. For the record, while in England and on the Continent IX TAC never received any U.S. beer.

France

On the afternoon of 23 August 1944 our U.S. Army Air Corps truck convoy edged its way slowly into the outskirts of Versailles, France. It had been a hard trip, and we were anxious to get our tent camp established as soon as possible. By 2 p.m. we reached Versailles Palace and started pitching our tents in the gardens between the palace and the Petit Trianon. A place which reeked with history was having some more thrust upon it. Louis XIV must have looked down, or up, on the scene with some disgust. Here a group of uncouth Americans was pitching tents and putting military field equipment next to the palace that had been the seat of his court. Here was where the peace treaty between the United States and Great Britain was signed in 1783. The city had been the German headquarters during the Franco-German wars in 1870 and 1871. Here was held the "peace" conference in 1919. Now here in 1944 the headquarters of the IX Tactical Air Command, operating in direct support of the First U.S. Army, was pitching its tents. The First Army-IX Tactical Air Command team had been the first on the beachhead in Normandy, and it was destined to become the first Allied army into Germany and the first across the Rhine. Give Patton and his Third U.S. Army all the credit he and they so justly deserve, but remember that Courtney Hodges and his First Army were there, too. And the First Army was helped in no small way by Pete Quesada, the brilliant commander of the air units that flew in combat in direct support of Hodges' troops.

Beyond our campsite was the Petit Trianon, the palace of Marie Antoinette, who had come in 1770 at the age of 15 from Vienna to Versailles to marry the young Duc de Berri, afterward Louis XVI of France.

We slept in pyramidal tents, four to a tent. Our office was a large storage tent which provided shelter for eight field desks and associated equipment. We had to pitch and strike our own tents. In France, Belgium, and Germany, as well as in England, we worked seven days a week. Sunday was no different from any other day, except that those who chose would be at chapel for an hour in the morning. At Versailles, by the next day, Sunday, our encampment was complete. All the tents were in place and MPs were at strategic positions so the camp was secure. Civilians, or anyone without proper identity, were not permitted in.

* * *

Paris hardly belongs to France. Paris and the rest of France are two different worlds. The world can forever be grateful to the German general who defied Hitler's order to destroy the city. Today Paris has lost much of its attraction for the casual tourist, but it has not lost its history, or its charm, for those with the knowledge or the inclination to get off the beaten tourist paths.

I first got into Paris on the day it was liberated from the Germans on 25 August 1944, and others in my group arrived shortly thereafter. There were sights that I shall never forget. The French girls had a delightful custom of never sitting on their skirts. If I were fortunate enough to have an appropriate seat at a night club, the girls getting up and sitting down at the next table could be as good as the show. There was no petrol for cars and the Metro was not operational, so the gals riding their bikes, not sitting on their skirts, provided exciting sights.

We had heard about "The Sphinx." When we asked natives for directions we drew either a complete blank or a big grin. Either we got questionable directions, or we did not understand those given. We finally got proper directions from a gendarme, and at about 10 a.m. we walked into a very plush lounge where some 20 girls were sitting at the bar completely in the nude. There were no other males present. We gulped.

* * *

At the Mums Champagne works at Rheims (pronounced somewhere between "Renz" and "Ronz"), I was conducted on a private tour through the winery and the cellars by the Frenchman who had made the first trip on the *Normandie* to ensure that the champagne was properly served. (I had asked for him by name, which had been given to me by a friend.) Among other things he showed me the cellars, which had been sealed off before the Germans occupied the city. At that time labels had been put on each bottle: "Welcome Our Allies." Confidence, eh?

The Germans did not discover those cellars, so when the Americans arrived there was a fair supply of champagne available, besides the normal production of the winery. But bottles were scarce. An American officer could purchase one bottle of champagne if he had a turn-in bottle. He was sold only one regardless of how many empties he had.

After the tour my host took me to a handsomely appointed café-like room, where we sipped champagne and ate cheese. I learned some things. The reason the second glass of champagne tastes better than the first is that upon contact with the air, a chemical reaction results which improves the taste. The proper way to judge champagne, if one is not in too formal company, is to take some on the top of the tongue, and draw air over it. Champagne is one wine that is good, and appropriate, for drinking at any time, e.g., before, during, and after dinner, breakfast, and so on. Never serve pickles, salad dressing, or vinegar in any form when serving champagne, as they dull the taste buds, and one does not sense the true flavor of the champagne.

The famous cathedral at Rheims, the former site of the coronation of French Kings, was unscathed.

* * *

For the first month or so after liberation, the people of Paris had it tougher than at any time during the German occupation, as far as material necessities of life were concerned. The reason was the unplanned rapid advance of the Americans. Supplies could not keep up with the front-line troops. Otherwise, the war might have been finished in September or October 1944, rather than in June 1945. If combat troops could not get sufficient supplies, it is not surprising that the people of Paris could not. At IX TAC Hq all we had to eat for 10 days was K-rations. This would not have been bad, but only the cheese ration was available. Cheese for breakfast, for lunch, and for dinner for 10 days. Had we not been able to liberate some champagne, we probably all would have died of constipation.

* * *

Our start home after hostilities ended was interesting. After Doc Taylor and I got to Paris we found that our flight to the U.S. had been delayed. What does one do after two years in Europe with an unexpected three days in Paris?

I probably would not have made it if our aircraft had left on time. When I checked into the American Embassy, they asked for my passport. I had none. They were horrified. A civilian in Europe without a passport! Unthinkable! I thought they were going to have me jailed, castrated, or guillotined. I said to the staff member quietly, "Mister, when I arrived on the beachhead in France not long after D-Day, I did not see a single member of the embassy asking to see my passport." To his credit, this pacified him, but I still had to get a passport, a picture for it, and the rest.

I met a Colonel Edwards, an Army doctor, and one night we went bar-crawling together. I do not know if we set a record. During the war the French had a saying to explain away anything that was not under their control, within their competence, or within their inclinations: "C'est la guerre!" (It's the war!) At one bar nothing was going right for the bartender. He was growling, edgy, flighty, and demonstrative. As he walked by me I said, "C'est la paix! ("It's the peace!") He went up and down the bar, repeating it and pointing at me. I had caused quite a sensation.

On my last night in Paris I went to l'Opéra Comique with Doc Taylor and Fred Seitz, now president of Roosevelt University in Chicago.

Brussels

It was a long time before we got a respite from front lines, foxholes, tents, canvas folding cots, shaving from our helmets, and showering under the apple trees with water from an oil drum overhead. Several of us got 24-hour passes for Brussels. Roland (Rolie) Larson and I decided to explore Brussels together. We stayed at the Metropole Hotel, which even by American standards was fine. White sheets! It had been a long time since we had slept on sheets. It turned out that Rolie and I did not sleep on those very long during the 24 hours.

After two showers each we decided to explore the city. We both liked beer. The only beer we had had since being on the Continent was French *bière*, which was hardly worthy of the appellation. Rolie also liked ice cream. I did not share his craving for it, but after we each had a beer at one establishment, I would sit with him while he had ice cream at another. Beer-beer, ice cream, and so on. I do not recall how long this continued.

That night we went to a night club, "Le Boeuf Sur le Toît" (the Bull on the Roof). A rather rough war was going on, and not too far from Brussels. Everything closed at 9 p.m. Outside the night club we saw two attractive girls, and, with my barely adequate French,

I started talking to them. The two went into a huddle and then motioned for us to follow them. Rolie and I looked at each other with smiles of conquest. The four of us got on a streetcar and rode to the end of the line, where they took us into a completely built-up area, buildings three and four stories tall. In the complete darkness we thought it was a warehouse or factory area. Finally the girls went to a door and rapped on it. We saw the cover on a peep-hole thrust aside.

We were taken up one flight of stairs to the very modern apartment of the parents of the two girls. Their mother and father spoke excellent English; the 12-year-old son did rather well. When asked where we were from Rolie said, "Chicago." Immediately the young boy braced his arms as if holding a machine gun and said, "Rat-tat, rat-a-tat, rat rat-a-tat!" The influence of Chicago gangster movies.

Soon the two girls went to bed and Rolie and I stayed talking with the parents and the son. Regardless of the subject, the son would interrupt with, "Rat-a-tat, rat-a-tat, rat-a-tat-tat." The father said that the Germans had a very effective way of enforcing blackouts. If they saw a streak or speck of light from any source, they would shoot through it.

Though he never revealed what it was, I am certain that our impromptu host had held some important role in manufacturing before the Germans took over. He lamented that those who had collaborated with the Germans were now well off, because their shops and factories were operational, turning out products for the Allies. He, and others like him who did not collaborate, had nothing, and no prospects. He thought this did not seem reasonable or equitable.

We finally left. The trolley on which we came had been the last one of the night, so we had to hoof it all the way into town. Moral: It pays to try to pick up pretty girls. Otherwise, we would not have had this very interesting experience.

Charleroi

Men under duress, stress, great danger, and losing close friends have to cut loose once in a while. But it was not a way of life. It happened infrequently. I had a theory as to when the war would end. If both sides ran out of liquor, it would be mandatory to call at least an armistice, so both sides could make, or get, some more.

For a while the Battle of the Bulge was a grave and uncertain time for the Americans. Hq IX TAC was at Verviers, Belgium, when it started, and the potential danger of the situation was indicated by orders to destroy all classified and official documents. We were ordered to Charleroi, by way of Holland, as the Allies did not know the location of many of the German units.

Upon arrival at Charleroi, IX TAC restricted all the personnel of its fighter units to base. Two IX TAC fighter units were at Florennes, about 20 miles south of Charleroi, and it was found that they had made elaborate arrangements for a New Year's Eve party in Charleroi. A large hall with two floors and two dance bands had been engaged, and many females had been invited. Since the two units were now restricted to their base at Florennes, Hq IX TAC did the most decent thing it could do. It took over the arrangements and the party.

Germany

For a Ph.D. in physics a knowledge of German and French (determined by examinations) is required. I had taken two years of high school French and had utilized my French outside the classroom, and this proved adequate to pass the French exam. But though Zimmerman is a good German name, I knew nothing about the language when I enrolled in a college German course. It was clear that my only requirement was to be able to read German. Speaking it was for the birds, or Germans, I thought. I did not have the same opportunity to learn German that I had French in France and Belgium. The reason was that there was a $65 fine for "fraternization" with a German.

We were able to liberate some German champagne. It was as good as "average" French champagne, with one exception: about every third bottle was completely flat. They needed to learn something about quality control.

We were located at Verviers during the siege of Aachen, Germany, about 15 miles away. We would occasionally go to a hill overlooking Aachen and observe our fighter-bombers attacking the city. Before it fell Aachen was quite extensively battered down. An early unmanned weapon was used by the GIs in Aachen. A streetcar on the outskirts was filled with explosives. A time fuze was set and then the car was dispatched on its own to the center of the city. When I got into Aachen, though all the nearby buildings were flattened level, a sign remained intact on a traffic circle. It quoted Hitler: "Give Me Five Years and You Will Not Recognize Germany." How right he was!

It was interesting to go through a narrow strip of former Germany between Belgium and Germany, which by plebiscite after World War I had become a part of Belgium. They had retained the German language. Thus the "Welcome Our Allies" signs were written, "Willkommen Unser Allierte."

* * *

Doc Taylor and I visited Buchenwald concentration camp shortly after its liberation. A former slave laborer who had been a captain in the Czech Army was our guide. He had believed the German propaganda distributed in Czechoslovakia, which essentially said, "Come to Germany for new opportunities in tech-

IN GERMANY DURING WORLD WAR II

Entrance to Buchenwald, Germany, May 1945.

Hq IX Tactical Air Command. L to R: Smeltzer, Eudey, Boffa, Zimmerman, Lettier, Larson.

nology and manufacturing.'' He, his wife, and his two children, with six trunks of personal belongings, took a train to Germany. Upon arrival, he was sent one way and his wife and children another. He did not realize then that he would never see them, or hear from them, again. He was taken as a slave laborer to Buchenwald.

At Buchenwald were huge ovens (or furnaces) where those who died of starvation or abuse were cremated. Outside the ovens were large hooks on the walls where live troublemakers were hung by some part of their anatomy, until they were thrown into the ovens. Thousands of slave laborers remained at Buchenwald on its liberation (the top population was estimated for us at 40,000), and the Germans tried to evacuate some by trains when the Americans were approaching. Thousands (estimated to us to be 5,000 to 7,000) were packed into boxcars and cattle cars. When the Germans learned that the rail line ahead was held by Americans, they machine-gunned those thousands to death. It is little wonder that some nationalities, when set free,

Zimmerman.

H. O. Wyckoff.

On left, Mark Eudey and L. S. Taylor.

were quite rough on Germans, civilian and military alike, whenever they had the opportunity.

One fenced-in area still had emaciated prisoners packed closely together. They were smiling, almost jubilant. Our uniforms clearly identified us as Americans, and one tall, extremely thin man yelled in perfect English, without an accent, "Where are you from?"

I replied, "Dayton, Ohio."

He beamed. "I'm from Cleveland."

He was born in Germany but as a child went to the U.S. In 1939 he decided to visit some relatives in Germany, and he was caught in the Hitler hysteria; he had been a slave laborer for years. He said he was going back to Cleveland at the first opportunity.

Even my generation does not have a complete understanding, feeling, and appreciation of the kinds of things I am revealing. So how can we expect our children to?

Buchenwald was where the infamous "Madam of Buchenwald" had lamps with shades made of human skin. We saw the lamps. Our Czech guide pointed to

Hq IX Bomber Command OA. Newmark, Bichowsky, Taylor.

Yost, Mohler, Wyckoff.

one building. It was where binoculars assembled there were stored prior to shipping. He said that the first GIs through took six to ten pairs each. If they had been satisfied with one or two, there would be one for each of us.

Our Czech captain repeatedly asked, "What are you going to do with these people (the Germans)? You cannot do the same to them that they did to us, but what are you going to do?" We had no answer. Even if we knew then that the Germans were to become one of our most important Allies, that answer would not have been satisfactory at the time.

No German living today (or for that matter immediately after the war) had ever been a Nazi, according to their assertions. Headquarters, U.S. First Army, and Headquarters, IX Tactical Air Command, moved into Weimar, not far from Buchenwald. General Courtney Hodges, the commanding general of the First Army, had signs posted all over the city, announcing that any male of any importance should appear at the city square at 9 a.m. the next day. Hundreds came, some with top hats and/or colorful sashes. They were loaded into trucks and hauled to Buchenwald, where they were directed to clean the place. Each German claimed that he had no idea such things were being done at Buchenwald. Many skeletal bodies were still in place, as well as thousands of emaciated living beings. Many Germans vomited; some fainted. Perhaps some really did not know what was happening there.

* * *

We were still in Weimar on V-E Day. An airstrip was nearby, and there was considerable excitement, as some German planes were landing there. They much preferred surrendering to the Americans than to the Russians. Our ack-ack would shoot a few rounds to keep them honest. The Germans would circle the field and then land. The first German pilot to land was unfortunate. After he got out of his aircraft, when he was reaching for his gun to give it to the American officers in surrender, a GI thought he had other ideas and shot him. He was not seriously wounded.

Several of us heard Lord Haw Haw's last radio broadcast. He was an American-speaking German propagandist who aimed at the British and the Americans news and propaganda to lower Allied morale and induce uncertainty. It was clear he had been drinking heavily. His plea: Americans, *now* turn against the Soviets. Blast them now, or you will forever be sorry you did not.

William Joyce, Lord Haw Haw, an Irish-American-British fascist, was hanged as a traitor in 1946.

American Push Through Germany

My father took a portion of a letter that I wrote in 1945 to my wife, from Germany, to the *Miamisburg (Ohio) News*. They printed it as follows:

> Experiences of a former Miamisburg man, with Headquarters IX Tactical Air Command in Germany, are interestingly told in a recent letter received by his wife, Mrs. Carroll L. Zimmerman, 819 East Maple

Avenue, who with her son, is residing here while her husband is in Europe.

Mr. Zimmerman writes: "I wish you could see the Rhineland in the Spring. With all the trees in blossom, with the castles perched on the high cliffs along the river, and the gentle rolling countryside — all this would make you wonder if there could be a war in such beautiful surroundings. But you never forget for a minute that there is. The planes in the sky, the endless supply convoys streaming towards the front, the battered-down German towns, the prisoners heading towards the rear, and the sullenness of the civilian population — all this keeps the issue quite clearly before you.

"It's a real pleasure to see German town after German town battered to the ground. All those houses still standing have a white flag of some kind flying from them — towels, bedsheets, and even undergarments. Hundreds of freed Poles, Russians, French, British, etc., making their way along the road as best they can. Some in uniform salute smartly as you pass. Small groups of German prisoners walk unescorted towards POW cages. A convoy of 20-30 big trucks packed full of German military has only one GI guard on the front truck and one on the last truck. German women weeping bitterly as the prisoners go through a town. Others waving with uncertain feelings towards them. German civilians hitting the dirt as an American plane swoops down to housetop level to have a look. Miles and miles and miles of American vehicles of all kinds in a steady stream across each of the many bridges now crossing the Rhine. Most of the bridges one-way now to speed things up. Many women and many 10-12 year old boys among the freed slave labor. A ration truck upset on a town corner with a civilian crowd looking on and wondering if they dare touch. . . .

"The biggest thrill I've had so far in this war was when we came onto the Remagen bridge, still up. Now its center spans lie in the river but its towers on either side still symbolize the drama that occurred when it was found usable. This was one big break we got in the war and how much it will shorten it we can only estimate. It very well could have been the turning point.

"Koblentz, at the entrance to the Saar, badly battered. Bad Godesburg and the hotel where Chamberlain had that fateful meeting with Hitler. Cologne, where the spires and main walls of the cathedral still stand in the shell of a city that once had 700,000 people. Bonn, with most of the city as well as the world-famous university in ruins. All these are found along the beautiful Rhine. A year ago I would have felt sorry about such things. Now I'm disappointed when I come to a town that isn't completely battered down. Aachen, Duren, and Munchen-Gladbach, the home of Goebbels, are towns that I have seen and which are practically destroyed. It certainly is difficult to understand why the Germans still resist."

Chapter 4

Other Impressions and Reflections

Bureaucracy

It is a wonder that the U.S. won the war, considering the paperwork and bureaucratic procedures required, even in the Combat Zone. Unfortunately, today too many bureaucratic procedures burden governmental operations. Two horrible examples from World War II are given here.

After I had been in Europe for a little over a year, the Pentagon recommended that my per diem be increased by $2. This took three different basic letters with a total of 13 endorsements, along with many months. In various stages the correspondence had to be handled by General Eisenhower's Hq four times. A good example of Parkinson's First Law — less trouble getting 1,000 B-17s than a $2-a-day pay raise. Copies of the complete correspondence are in my files: it lasted from 16 July 1944, when the raise was first recommended, to 20 November, when Hq Ninth Air Force acknowledged that the "claim appears to be in order." I cannot remember when, but eventually I received the $2 per diem retroactive to 16 July.

Near the end of my two years in Europe someone at higher headquarters decided that we civilians should be paying 50 cents a night for living in a tent, $1 a night if in a structure of some kind or another with two or more in the same "room," and $2 a night if in a structure or room alone. Can you imagine paying 50 cents a night for living in a tent with three others in the mud in the Combat Zone within a few miles of the front line? While being shot at and bombed, and when you were not drafted, but a volunteer? The Operations Analysis civilians at Ninth Air Force headquarters

Awards given by Maj. Gen. St. Clair Streett, Commander, Continental Air Forces, 1946. L to R: Zimmerman, Brig. Gen. Fred Smith, XXX, General Streett.

paid (I don't think they ever stayed in tents, as their Hq was never close to the front line). The finance officer at IX TAC completely ignored the issue. He had had enough with retroactive pay adjustments.

* * *

After the war, while General Eisenhower was still an Army general, I heard him give a talk at the National War College. I cannot remember all that he said, but one thing stays vividly with me. He said something like, "Don't waste your time in trying to change organizations. Only one in a hundred of you would be effective. Instead, understand the organization you are dealing with. Take the necessary actions and routes under those circumstances that you must (regardless of whether they seem sensible or necessary to you) to accomplish what you think is important."

Military Maxims

After World War II General Gruenther, top Department of Defense man in Paris, was lamenting that civilians were so deeply involved and influential in military strategy. Once he stated it something like the following: There are only two situations where the amateur is better than the professional. And the second of these is in military strategy.

In office at Bolling Field with Modie and Carroll, Jr., after being awarded the Medal of Freedom, the highest award the nation can give a civilian. Zim, Jr., wears it.

In 1942-1943 I was doing research in radar in the Airborne Division at MIT. We had many military visitors. One day, an Army major general watched the performance of a radar at East Boston Airport tracking a live aircraft flying above. He turned to me and asked, "How does it operate? On sound?"

Navy officers were, at that time, much more knowledgeable and interested. They would discuss operational considerations with us — displays, operational concepts, and so on. I wondered why this was so, and finally decided that it must be because when the ship sinks, the admiral has to swim, too.

I was walking down a street in Belgium with Brigadier General Kincaid. I was in uniform without rank insignia. As we passed the garage that had been taken over as our motor pool, a GI slouching against the wall didn't move. General Kincaid stopped and said, "Soldier, why didn't you salute?" The GI, now rigid at attention, said, "Oh, excuse me, Sir. I thought you were a lieutenant colonel."

Some classic tales that I think merit repeating are:

When the Germans completely surrounded the Americans in Bastogne, a GI said, "The poor bastards. They have us surrounded."

When the Germans sent their surrender ultimatum, then Brig. Gen. Anthony McAuliffe replied, "Nuts!" This helped the American cause considerably because those Germans assigned to try to figure out what that meant were not available for combat. Bastogne did not fall.

A major was briefing a brigadier general in his office at SAC headquarters. The general stopped the major at one point and said, "Either I am stupid, or I do not understand the English language." The major replied, "Are those my only choices, Sir?"

Prior to World War II Brigadier General Kennedy had received some psychiatric attention. He was given a clean bill of health. During the war he sometimes would disagree with some of his colleagues over their planning responsibilities. Occasionally when he thought an important point of his was not having any impact, he would take a card from his wallet, lay it on the table, and say, "I can prove I'm not crazy. Can you?"

Lt. Gen. Arthur C. "Sailor" Agan (USAF Ret) told me the following experience. He had been a test pilot at Wright Field, and on the morning of a scheduled test flight, the weather looked poor to him. A test pilot has enough hazards without having weather hazards imposed on him. The weather was posted as the minimum permitting flight. Let us say the required visibility was half a mile. He had the audacity to ask the weather officer how they knew the visibility was half a mile. The officer said, "Son, come here." He went to the window and pointed in a certain direction. "There is a tower exactly five-eighths of a mile away, and I just can't quite see it."

That overtraining is possible was demonstrated in a story General Power told me. During the war, while commanding a B-24 Group in Italy, he was watching his B-24s take off. One didn't make it and slithered off the top of some trees. This reduced the crash impact so that all crew members were able to get out. The B-24 was considerate enough not to explode until the crew got a safe distance away. A total loss in flames. General Power asked the young captain aircraft commander how many inches he was pulling. The lad replied, "Thirty-nine."

"Why didn't you shove the throttles against the fire wall?"

"If I had done that I might have ruined the engines."

Chapter 5

Headquarters, Continental Air Forces, and Strategic Air Command

Chief, Operations Analysis

In 1945 I returned to the United States in June, after over two years in Europe. After several weeks of leave in Ohio I reported as Chief, Operations Analysis, Hq Continental Air Forces, Bolling Field. (When the Air Force became autonomous on 26 July 1947, a base was designated as an Air Force Base [AFB]. Hence Bolling AFB. Prior to that, Army Air Corps bases were known as fields, hence Bolling Field.)

Hq Continental Air Forces (CAF) was over the four training air forces in the ZI. I reported to Maj. Gen. St. Clair (Bill) Streett, Commander. Maj. Gen. Monk Hunter was commander of the First Air Force in Florida, under CAF.

A message was received from Hunter quoting a regulation that stated that the senior officer in the rank required for a particular position should be commander. He was senior to Streett. Therefore he should be Commander, CAF. General Hap Arnold handled this quite adroitly, as he did so many things. He put out an order giving himself the additional duty of Commander, CAF, with Streett as deputy commander. Of course, General Streett continued to run the show.

It was quite a challenge to be Chief, Operations Analysis, in a command whose main mission was the separation of its officers and men. We studied aircraft accidents. We proved accident proneness. Our efforts were partially responsible for establishing the USAF flight safety organization at Norton AFB, California.

A special photographic unit had been established during World War II at Lookout Mountain, California. As I recall, this was the unit Ronald Reagan was assigned to during the war. There were thoughts of disbanding the unit. General Streett asked me to prepare a case for keeping the unit together and assigning it to Streett's command. Several weeks after the recommendation was sent to Hq USAF a reply was received. General Streett brought it into my office. Hq USAF had decided to keep the unit together and assign it to the Air Research and Development Command (ARDC).

Maj. Gen. St. Clair (Bill) Streett, Commander, Continental Air Forces, 1945.

First Commander

The Strategic Air Command was established on 21 March 1946 at Bolling Field. Later Hq SAC moved to Andrews Field, Maryland, and then to Offutt AFB, Omaha. I joined Hq SAC on the day it was established, as Chief, Operations Analysis.

I reported to General George C. Kenney, the first commander of SAC. Major General Streett was deputy commander. General Kenney was set upon becoming

Chief of Staff, United States Air Force. I knew this was not going to happen several days before the staff did, as I had seen a message which stated that no longer was the senior officer in grade automatically commander. The requirement was that any named commander must have the grade required by the position. That resulted in a general or flag officer being promoted from two stars to four stars and then being named chief of his service (not in the Air Force). The SAC adjutant general did not immediately distribute the new regulation. I do not know if this was inefficiency or apprehension. General Hoyt Vandenberg became Chief of Staff, USAF, in July 1948. General Kenney must have known.

U.S. Air Force Arctic Training Films

When SAC was still at Bolling AFB a major came to me one day and asked if I had any suggestions for training films; he had a letter saying money was available from the Pentagon. I certainly did. My office had been given the responsiblity for the analysis of SAC Arctic and Polar flights. When I tell you that in one week the 72d Reconnaissance Squadron did more Arctic flying than had been done in history, you will realize that data were available to analyze. Later my office was given the responsibility for analyzing all of the Army Air Corps Arctic flying activity.

The letter had stayed in the major's desk for 30 days, and if money was to be available the proposals, with technical scripts, had to be in the Pentagon the next morning. One of my analysts, a master sergeant, a secretary and I worked all night. We got the job accomplished, and we got the money.

A Hollywood film company was contracted to write the professional scripts in cooperation with my technical script writers. They did the filming and the directing, but Army Air Corps personnel were the actors. The film company had bee doing MGM specialties, like the Pete Smith show and Robert Benchley. This group broke from MGM and formed their own company. Jack Chertok was president and Harry Poppe was producer. Later they made "The Lone Ranger," "Sky King," "Private Secretary," and others.

I looked at every Arctic training film I could find, and read anything I could get my hands on concerning the subject. Among other things I insisted that the filming be done on location, in the Artic and Polar areas. I had seen one Army training film in which the temperature was said to be 40° below zero, yet a GI picked up a steel stake in his bare hand. My insistence turned out to be a challenge. It's one thing to show how an aircraft crew of six can survive on the Polar ice cap, and another thing to have 30 others survive while doing the directing, filming, etc. Also I insisted that the films be made in color. I can still hear the Pentagon general saying, "Why in hell do you need color to take pictures of ice and snow?" But I prevailed.

Five official survival films were made, each over an hour. A member of my office was technical director of each film, on location of course. The program took over two years to complete. I was producer. At one time, in addition to the Hollywood contract personnel, I had six or seven aircraft and over 100 Army Air Corps personnel on location in Alaska, Greenland, and the Canadian Archipelago.

Unknown to me, The Air Corps entered one of the five films, "Survival on the Arctic Tundra," in the international film festival at Cannes, France. It won first place in international competition in the Documentary category.

Thule, Greenland

Greenland is a Danish possession. Thule is on the west coast of northern Greenland about 900 miles from the North Pole. Our analysis of all of the Army Air Corps Arctic and Polar flying included SAC's reconnaissance squadrons, the air weather service, and others. Some of our men became experts by assignment. Much study and personal participation made several of them true experts. At two different times two of them, Bob Davis and Wally Boffa, were at the South Pole.

Bob Davis landed a small aircraft, alone, on a North Polar ice pack to see if one could survive and then take off from the ice floe. I had photographed, from the air, areas of that part of the world for the first time in history. I put aircraft and men on the surface for the first time in history in some areas.

I had other expert help. Bernt Balchen, the famous Norwegian polar explorer, joined the U.S. Army Air Corps as a colonel early in the war. He was a personal friend and a consultant to me for years. He has since passed away. Shortly after the war Bernt suggested that I hire Hans Siewers, a Norwegian who had been hunting in Greenland when World War II started and stranded him. For the duration he had helped rescue Army Air Corps crews off the Greenland ice cap. Remember, there was a considerable ferry operation over that area to Europe, some 1,300-1,500 aircraft.

I got in touch with Hans and hired him. One of his first assignments was to survey sites for a possible air base in northern Greenland. After much study, he went alone to Greenland and hired a dog team and sled commanded by a son of Admiral Robert E. Peary, who "discovered" the North Pole in 1909. Hans visited and surveyed some sites, and then chose Thule, a small Eskimo village. In 1950 the Joint Chiefs of Staff accepted Thule as an airbase site. Sometime after Hans got back, I received a formal invitation to the Danish Embassy. I was told that the next time I sent someone into their territory it would be nice if I would inform them ahead of time.

Thule Air Force Base has all its pipes and sewers insulated and heated. There are tunnels to move among the various buildings in winter. Poles carrying wires to remote radar sites were tied down with many rocks bound to their bases, to enable them to withstand the winds. A small bar at one of the remote sites used ice a million years old in their drinks: they cut it from the base of the ice cap.

When I was flying over the Greenland ice cap in a KC-97 en route from Europe to Thule, General LeMay called me on the aircraft from Omaha to tell me that Rod Smith's and Stan Lawwill's first supergrades had been approved by Washington. And I got a particular thrill sitting on the ground in a KC-97 at Thule and talking to Little America, South Pole, via single sideband.

A Ballistic Missile Early Warning System (BMEWS) site was just outside Thule. Antennas larger than football fields. Miles of wave guides big enough to walk through. Klystrons as large as a man. My first impression was that an electronics type better never talk to me about the size of bomber aircraft. If he did, I could engage him in discussion.

A considerate Secretary of the Air Force, Dudley C. Sharp, kept me from being stranded at Thule on one trip. After a visit to the BMEWS we were scheduled to take off for Washington. A car and driver and escort officer had been assigned to me. My officer didn't get the word. He took me back to my quarters, whereas everyone else went directly to the aircraft. When we comprehended the situation we went to the flight line as fast as we could. The Secretary and everyone else was aboard, patiently, or otherwise, waiting. The moment I stepped aboard the aircraft commander started the engines.

Maj. Gen. Clements McMullen

Maj. Gen. Clements McMullen replaced Maj. Gen. St. Clair Streett on 10 January 1947 as deputy commander of SAC. For the first months he ran the show, as General Kenney, the first SAC commander, spent most of his time completing his responsibilities at the United Nations. McMullen tried hard to streamline SAC and increase its effectiveness. However, he emphasized some of the wrong things and some of his methods were so extreme they backfired.

If a member of Hq SAC drove through the gate later than eight minutes to eight in the morning, the MP would take his name and hand him a mimeographed note signed by McMullen saying that if one was not through the gate by that time, one could not be at his desk by eight. After two or three of these it was necessary to see McMullen personally. The result? If one could not get through the gate by eight to eight he waited until nine a.m. after which the MPs did not stop cars and kept no records.

McMullen gave the staff two working days to reply to an electrical message and five working days for written ones, regardless of the subject's importance. To his credit he thought that if a final reply could not be prepared, an interim reply should be dispatched. But he had the adjutant general keep score and each Monday in the lobby of the new building was posted the delinquent-action status of each office, by name, for all the world to see.

He did effect some efficiencies in manpower in SAC.

He was very helpful to me in getting my office responsibilities defined for many on the staff, who had never heard of Operations Analysis, and he gave me important responsibilities for all SAC requirements actions. General LeMay kept and increased these when he took over. McMullen told me that civil servants at the lower levels were paid too much, and at the higher levels much too little.

Under-Secretary of War Royall decreed that in the Army and the Army Air Corps 40 percent of all civil service positions GS-5 and below, and 60 percent of GS-6 and above, would be eliminated. McMullen dispatched correspondence which successfully excluded Operations Analysis, SAC, from the decree. Note I said Operations Analysis, SAC, and not Operations Analysis, USAF. McMullen's letter started, "I have 'n' operations analysts and I intend to keep them."

Most of us have seen that when a military man is downgraded, deservedly or unjustly, he often bounces back if he buckles down to his degrading job. This happened to then Brig. Gen. Fred Smith, who was Chief of Staff, SAC. He thought it was his option and responsibility to get his two-bits in, though if the decision went against him, he would support the official decision. His views often disagreed with McMullen's, who had him ostracized. He was made commander of the Civil Air Patrol, a double insult. A general officer had never been commander before, nor a regular officer; he had always been a reserve officer.

About six months later Smith and I were at the same cocktail party. Freddie said, "Zim, I have an eager beaver for a deputy. He tries to go so many directions at once, that he goes nowhere. He thinks he should have my job, and so do I."

Freddie bounced back. He retired with four stars, with his last assignment being Vice Chief of Staff, USAF. McMullen retired with two stars.

General Curtis E. LeMay

Before I met General LeMay I smoked cigars and signed my first name, Carroll, with an old English "C." He signs Curtis the same way. Hence neither of these were affectations on my part.

I first met General LeMay when he was Deputy Chief of Staff, Research and Development, at Hq USAF. I had recruited Dr. Scott Rethorst from MIT as a member of my staff. Rethorst had an idea about a unique propulsion system, and another MIT professor, Dr. Raucher, was interested. Raucher knew of some research in Europe along the lines of the idea, so I proposed hiring him as a consultant and sending him and Rethorst to investigate the research abroad. Col. Don Putt at Wright-Pat AFB refused the request. Since it was Research and Development I had coordinated the request with Wright-Pat.

I thought it was important, and I was able to schedule a meeting with the Deputy Chief of Staff, Research and Development, at Hq USAF, who was General LeMay. I explained our purpose to him and told him of Don Putt's nonconcurrence. General LeMay approved the trip, which turned out to be beneficial.

(As Don Putt advanced through his career, I had quite a satisfactory association with him frequently on issues and projects.)

Lieutenant General LeMay assumed command of SAC on 19 October 1948 at Andrews AFB. (Prior to this, the decision had been made to move to Offutt AFB, Omaha; General LeMay had had nothing to do with the decision.) SAC became operational at Offutt on 9 November 1948.

* * *

During the first week after General LeMay became commander he called me to his office. I briefed him, no charts, on what we were doing. He said he wanted me to really understand the Strategic Air Command, to go to briefings and staff meetings of the director of Plans and the director of Operations. I should look at his schedule each morning and join him in any part of it I wished.

Whenever I wanted to see him, I was to come in, even if his door was closed. If he didn't want me there he would throw me out.

He would fire me if he caught me pushing too many papers. My main job was to understand SAC, so I could better assign chores to my people and represent him at pertinent meetings at lateral and higher headquarters.

I took him at his word and was never thrown out of his office. The Chief of Staff, SAC, Maj. Gen. Richard Montgomery, would occasionally ask me, when he knew I was going to see General LeMay, if I would ask him this or that. Frequently his aide would phone me and say, "Hey, the Boss is restless. Can you come and talk with him?" I always could.

I regularly scheduled briefings for him, and for General Power, on work we were doing. Often it was not a finished product, and was not asking for a command decision. This provided an opportunity for us to get his comments, which were never prejudiced or directive but were often, "What would happen if you assumed this?" "What would happen if you did that?"

Those questions were often the key to the analysis, and after a while one can ask oneself the same kind of questions. I have briefed the Secretary of Defense, the Chief of Staff of the Air Force, the Joint Chiefs of Staff, Congressional leaders, the chairman of the Atomic Energy Commission, and others, when the first time General LeMay heard what I was going to say was at the briefings.

Awards given by General LeMay, CINC-SAC, to Stan Lawwill and Zimmerman, Hq SAC, 1958.

Celebrating 10 years of service as civilians, Hq SAC, late 1950s. Back row L to R: Kingsbury (MAT), XXX, Chittim (OA), XXX, Bohn (Historian), Zimmerman (OA), Boffa (OA), Balmer (OA), Lawwill (OA), Harding (P&A), McKenna (P&A).

Omaha, Nebraska

We moved to Offutt AFB, Omaha, when SAC Hq moved there in early November 1948. We left Omaha in June 1959 for Washington, D.C., and Hq USAF.

I had heard of foreign people getting only the word that the authorities wished them to get. I really did not know it existed in the U.S. until I moved to Omaha. There the two newspapers had the same owner. No TV in 1948. Radio news left much to be desired. One could subscribe to the Congressional Record, but how many would at $45 a year? And who would know what it was saying? To be fair, Omaha was not denied news. It was simply that it was not given some, and inertia is a powerful force.

When the move to Omaha was finally decided upon, Stuart Symington, then Secretary of the Air Force, went to Omaha and gave a speech to the Chamber of Commerce. Among other things he said that now he hoped to be able to get some work done, that Senator Wherry had been in his office for the last six months.

Prior to this SAC Hq had been scheduled to move to Colorado Springs. I had my orders in hand. Being on the ball, I had gone there and purchased a home. The day before the van was to pick up my furniture the move was canceled. Others were not as fortunate as I; they had their furniture there or on the way. Other candidates with Omaha had been Miamisburg, Ohio; Indianapolis, Indiana; Scott Field (in Illinois near St. Louis); and Shreveport, Louisiana.

Carroll Zimmerman at Hq SAC, 1950s.

* * *

The city responded in a magnificent fashion to a true emergency. The Missouri River was at its highest flood stage in recorded history. If the dikes broke on the Council Bluffs, Iowa, side Omaha would be saved. Many of the people in Council Bluffs were evacuated. If the dikes broke on the Omaha side Council Bluffs

General LeMay with Arthur Storz, Air Force Association official.

At Offutt AFB, 1950s, with Modie.

would be saved, and downtown Omaha would be under 16 feet of water. Volunteers filled, hauled, and carried sandbags to the dikes; they were built four to six feet higher. A dozen professional men from my office worked around the clock, though their homes were not at hazard.

Teenagers drove large trucks with sandbags, cement, railroad ties, rails, and paid no attention to traffic lights. Police visited the schools and all the homes they could, asking that children be kept off the streets, pointing out that traffic lights would not be observed by the trucks. The commercial airport closed — only a dike between it and the raging river. Here the battle was almost lost. A sand boil developed, but it was spotted immediately. Sandbags, cement, railroad rails, and everything of substance available were thrown into the boil. The danger was conquered.

Aircraft were landing and taking off around the clock

At Offutt AFB, 1950s, from left: Nebraska Governor and Mrs. Val Petersen; Lyle Remde, Head, Air Force Association district; Mrs. Power and General Thomas Power; Modie and Carroll Zimmerman.

At Offutt AFB, 1950s, Edward Teller, Rodney Smith, Zimmerman.

at Offutt AFB, the site of SAC Hq. Troops, food, lanterns, tents, bedding arrived in magnificent numbers. It was a very heartwarming experience to see how the people could, and would, respond to an emergency. The people won. No fatalities and only a few injuries, which was remarkable. The people of Omaha showed their mettle in a dramatic fashion. This is the hope of America.

Dr. Edward Teller visited me during this time. As he stood on the dike watching the mad, swirling Missouri River, he calculated in his head how much higher the water was at the center of the river than at the level along the dikes.

After having rented a house for 10 years in Omaha and having twice refused General LeMay's requests to join him in Washington, I bought a house from General Gruenther's sister. Five months after buying the house we moved to Washington, D.C.

At Offutt AFB, 1950s, L to R: Wen Dwyer, Rod Smith, Zimmerman, Stan Lawwill, Paul Hower.

Chapter 6

Some Operations Analysis Contributions to Early Strategic Air Command

Stanley J. Lawwill

It is dangerous to single out one man from a group of outstanding men. The danger is that the others may be slighted. However, there is no doubt that during my 13½ years with Hq SAC, Stanley J. Lawwill was the most outstanding member of my staff. He was with me for 11 years at SAC as a division chief, deputy to me as Chief, Operations Analysis, and deputy to me when I was appointed Chief Scientist, SAC.

His academic degree was in mathematics, but he became a renowned operations analyst, nationally known. He had intelligence, initiative, industry, imagination, and ingenuity. He got along very well with the military. He appreciated the people who worked for him, gave them opportunities, and gave them credit. He was intensely loyal. His dedication was often at personal sacrifice, and at personal risk.

Among others, Stan has two unique accomplishments: he probably is the only U.S. citizen to have mumps on Eniwetok. He had been sent there for several months of temporary duty with the Pacific Proving Grounds. And the birth announcements for one of his children were sent from Eniwetok. In Omaha his wife gave them to him to mail. He put them in the pocket of his raincoat. After he got to Eniwetok on another assignment, in a downpour he put on his raincoat, and discovered the birth announcements.

Paul Hower.

General LeMay's Approach

General LeMay had the ability to select competent people, convince them to join the organization, give them important things to do, expect them to do those things well, and give them enough elbow room to do their best. Part of that, or probably the result of it, was a truly professional environment.

General LeMay had an outstanding ability to assimilate all the available important information about a problem, cut through the critical elements to reach a decision, and then get that decision implemented. I remember many meetings in the command conference room and in his office that demonstrated that ability. After briefings, during which he characteristically listened without interrupting, he would ask a question or two that went to the heart of the subject and frequently almost served as a brief summary. He then stated his decision so concisely and clearly that there could not be any question in an intelligent person's mind about what he needed to do. In those days it was possible to go ahead and get things done expeditiously.

Thomas W. Chappelle

A situation which illustrates General LeMay's confidence in his team, and his expectations of per-

OPERATIONS ANALYSTS, OFFUTT AFB, NEBRASKA, 1950s.

Ed Berdahl.

Wally Boffa.

Harold Balmer.

Harry Chittim.

formance from it, involves Tom Chappelle. It happened in the few months after Tom joined Operations Analysis, before he became my deputy. In those days we manned the office on Saturday mornings with skeletal staff. A division chief manned my desk to be immediately available to my intercom and its connection to the Command Section, on those rare occasions when I wasn't there. On Tom's first Saturday in that position he was working on a problem when the intercom signaled a call from General LeMay. He pressed the proper key and said something like, "Yes, sir. Chappelle here." At that time Tom was at most no more than a name to the general. Nevertheless, on the assumption that anyone I would entrust to the office had to be competent, he immediately said something like, "Chappelle, several of us are sitting here trying to decide whether we should undertake Project Reliable. As soon as possible, and in any event before the morning is over, I need to know the availability of B-36s per month if we undertake Reliable, and if we don't."

At that time Tom may have heard the name Project RELIABLE, but he knew nothing about it. However, from what General LeMay had said, he judged that any work Operations Analysis might have done on it would have been in Wen Dwyer's division. Fortunately, when Tom telephoned, Wen was home and could relate to Tom the essetial facts about Reliable and tell him that the basic data he needed for a quick analysis was available in Materiel. Again fortunately, Colonel Dickson was on duty. Tom got the needed data, analyzed them, derived the aggregate numbers General LeMay needed, and reported them to him via the intercom. General LeMay thanked him. SAC did implement RELIABLE.

Another situation that illustrates General LeMay's insight was the creation of the Office of Chief Scientist. I had asked Tom Chappelle to recommend arrangements to improve even further the service we were providing to the command while enhancing the stature of the science function in the command. At that time the Operations Analysis office was having many contacts with senior scientists and executives in other organizations, often as representatives of the commander. The recommendation that Tom developed was to create the Office of Chief Scientist, to consist of me, one other professional, and a secretary. My deputy and I would carry the major part of the load of representation on the outside, leaving Operations Analysis freer to concentrate on analyses, studies, and scientific projects.

I authorized Tom to recommend the reorganization to General LeMay during my absence from Omaha on a trip. Tom explained the recommendation to General LeMay in his office — no briefing, and eyeball-to-eyeball discussion with the general at his desk and Tom in a chair at the end of the desk. General LeMay listened silently, thought for a moment, rotated his chair so he could reach his intercom, pressed the key for General McConnell, Director of Plans, and said something like, "Mac, Chappelle is coming right over to your office to tell you about something I want implemented at once."

Tom went directly to General McConnell's office and explained the proposal to him. He rotated his chair to his intercom and informed both the director of Organization and Management and the director of Personnel that Tom would be in their offices within minutes and they should take the actions necessary to implement the changes he would explain to them. Within an hour from the time Tom entered General LeMay's office, all the actions to effect the changes had been initiated. Try to accomplish something like that today!

One important result of General LeMay's reliance on the people he selected for senior positions was that he had time to be available on short notice. By concentrating on basic strategies and major decisions, while depending on his staff to formulate them, he escaped the trap of a bulging schedule that would have made mature planning difficult. As a result, he was able to stay in complete control of SAC's operations, while being one of the most available persons in the headquarters.

Bruce Stafford

Bruce Stafford came to SAC via the personal contact route, having worked with Tom Chappelle at the Naval Research Laboratory evaluating the performance of naval anti-aircraft gun/missile fire control systems. In particular, he conducted some electronic countermeasures (ECM) tests of SAC B-50s at the NRL field station on Chesapeake Bay. The results of those tests (B-50 ECM effects on naval radars and fire control systems) he reported in NRL reports. They were of considerable interest to General LeMay and SAC (1952-1953 as I recall).

When Bruce joined Paul Hower's division in Operations Analysis at SAC Hq in early 1954, his first substantial project was to prepare a briefing for General LeMay describing electronic and infrared countermeasures — basic concepts, equipment functions, effects on SAC aircraft systems, and so on. After considerable planning, many fact-finding expeditions, hours of chart-building, and many dry runs, Bruce briefed General LeMay for an uninterrupted several hours. His purpose was to impart everything he knew about the subject to "the Boss."

Bruce describes the experience as wonderful, and a great way to get initiated into the SAC team. He learned firsthand the close cooperation and mutual respect that existed between the military and the civilian operations analysts. General LeMay was attentive, inquisitive, and understanding (all the qual-

OPERATIONS ANALYSTS, OFFUTT AFB, NEBRASKA, 1950s.

Wen Dwyer.

Max Oldham.

Gene Jackson.

Fred Fennema.

ities a briefer likes to encounter). Bruce's pre-meeting anxieties had quickly disappeared.

Bruce's involvement with the staff, principally Operations and Intelligence, became intimate and concentrated. General LeMay's interest in the penetration problem led to the formation of the "Reddoch Committee." Col. Joe Reddoch was then deputy to the director of Plans, Maj. Gen. J. P. McConnell. The charge to the committee was to find out what was going on in all phases of penetration aids (pen-aids) for the manned strategic bomber — research, development, production, operational use, testing, countermeasures, and other elements. The findings and recommendations for enhancing penetration capabilities of the SAC bomber force were to be reported to General LeMay.

Gene Jackson and Bruce lived with a group of 6 to 12 officers from other parts of the SAC staff, including the 376th Bomb Wing and the 544th Reconnaissance Technical Wing, on the road for several months. Using the aircraft assigned, they visited Air Force installations, contractors, "think-tank" organizations, universities, and almost anyone who was connected with any phase of bomber penetration.

This was done in early 1955, and Bruce recalls a few "heavy landings," many nights of burning the midnight oil, talking to many sharp individuals, working with dedicated and competent officers, and a general feeling of having done their best when the briefing was finally prepared and given to the "old man." This effort was partly responsible for speeding up the pen-aid development programs and redirecting some of them so they would be of more benefit to the SAC force.

When Gene Jackson and Bruce rejoined the Operations Analysis group after the Reddoch Committee work, they found that Hower, Camp, Burnight, and Cleaves were also deeply committed to the penetration problem. Bruce began to work more into modeling — from fighter-bomber duels to large SAC strike forces penetrating enemy defenses to strategic targets. They developed and applied these models to lay on strikes, and to evaluate the results of Emergency War Planning. This was long before the big electronic computers were available. Bruce describes this as quite an experience for a farm boy from Nebraska — to be working with military officers, shoulder-to-shoulder, figuring how best to destroy targets on the other side of the world.

A large project, in terms of manpower and importance, concerned with penetration aids for the SAC bomber force, was the joint SAC-Rand study conducted at the Rand facility in Santa Monica during most of 1956. Bruce was assigned full time to work with Rand in designing and getting operational the war gaming models on their computers. SAC Operations officers worked on the design of various tactics, and SAC Intelligence officers insured that "proper" enemy defenses were included.

The effectiveness of various bomber penetration aids, in being, being developed, or being considered for development, was determined for a variety of conditions. Paul Hower, Operations Analysis, and General Butch Blanchard, then deputy director of Operations at SAC, were jointly directing the SAC effort. Col. Jack Catton reported to SAC Hq at the time to be chief of the Requirements Division. He became the principal briefer, to give a military flavor. The results were widely briefed, including the Joint Chiefs of Staff, and they had a major effect on the pen-aid programs being considered at that time.

Not everything was related to the current war plan. Paul Hower started Bruce on improving the mathematical models to accommodate more sophisticated pen-aids, defenses, tactics, and so on, for the future. We developed a close working relationship with the Institute for Air Weapons Research at the University of Chicago. My people developed a set of models that could be applied to a variety of SAC penetration problems, both present and future.

These models provided the basis for war-game analysis many years later on the large computers used by the Joint Strategic Target Planning Staff (JSTPS) for Emergency War Order evaluation, and in Plans for force structure studies. At the time we were doing things manually. Bruce has vivid memories (as do many of us) of running the future "1959 War Game" for the director of Plans in our office with a series of overlays on a map, depicting the entire target system and enemy order of battle, with paper, pencils of various colors, slide rules, and a small electric calculator.

General LeMay and General Power often participated personally in these early war games. The results were of minor importance to them, but they were fascinated by the myriad decisions that had to be made with changing situations in the progress of the game. It also made it difficult for any Whiz Kid to give them a "snow job" concerning a war game.

The record must be set straight on another matter. In many quarters SAC was wrongly and ill-deservedly accused of planning for city destruction and mass civilian killing. The JSTPS developed the strategic target system. The Commander-in-Chief, SAC, CINCSAC, was the director of the JSTPS. The deputy director was a Navy admiral, who worked full time at the JSTPS facilities in the Hq SAC building at Offutt AFB, Nebraska. The staff, also full-time at Offutt, consisted of Air Force, Navy, Army, and Marine personnel. Their recommendations were given for approval to the Joint Chiefs of Staff and the Secretary of Defense.

This mistaken notion concerning target systems is a good example of the Zimmerman-generated SAC

OPERATIONS ANALYSTS, OFFUTT AFB, NEBRASKA, 1950s.

Stan Lawwill.

Rod Smith.

Adage: The expertness on SAC of an individual is directly proportional to the distance he is from SAC. Those farthest away are the greatest experts.

Edgar Berdahl

Ed Berdahl joined Operations Analysis, Hq SAC, in May 1950. He recalls that he was pleasantly surprised to find such a stimulating group of scientists of varying backgrounds in Operations Analysis. There were alert and competent officers of all ranks coming to the office to consult with our analysts.

Part of Ed's orientation was a trip to an operational bomb wing, the 97th Medium Bomb Wing at Carswell AFB, Texas. He was escorted there by Harry Chittim, who was an expert in dealing with all ranks and had a great rapport with them all. Ed was impressed that they first went to General Hutchinson's office to inform him why they were there. Col. R. M. Montgomery was Chief of Staff, and he made sure they had cooperation at all levels. The welcome mat was out to Ed, a novice, and he was impressed with the way the civilian operations analysts were accepted by their military associates. He was given a ride on a B-50 Radar Bomb Scoring (RBS) training mission. It was an outstanding orientation trip for a new operations analyst. It became evident that Colonel Montgomery, who had worked for General LeMay previously, was reflecting the CINC's desire that operations analysts should be treated as full-fledged members of the SAC team. Ed stated, "We were treated like they really appreciated us."

The next morning was a memorable one for Ed. As he and Harry left their quarters they saw that all the troops were wearing helmets and carrying rifles — the Korean War had started. As SAC was the command with atomic weapons capability, things became quite active and a bit tense.

Operations analysts most thoroughly understood the capabilities and limitations of atomic weapons, and their effects. When SAC Command Elements were formed, to be deployed forward into the Far East, North Africa, or European theaters, they included two operations analysts in the group of selected officers. Operations analysts would determine the optimum fuze settings for type of weapon, weather, delivery, and target conditions. Ed was quickly trained and became a member of the team that would be deployed to North Africa if such were required.

Ed recalls that he soon concluded that I had a special relationship with General LeMay and General Power, and that I had people working on the big problems which truly concerned the Commander-in-Chief and his key staff.

As a junior member of the office Ed was assigned to help more senior analysts, such as Rodney Smith and Wendall Dwyer. He assisted Rod in defining the necessary range capability of a truly intercontinental strategic bomber, when pre-strike and post-strike base structure was considered. The work was done for Col. Sherman Wilkins in Operations Plans and Col. Ariel Nielsen in Requirements.

Next came the missile age. Ed had a keen interest in the area, and he was going to planning meetings, project status reports on the early liquid-fueled air-to-surface missile RASCAL, and the ground-to-ground cruise missile SNARK.

In the early 1950s Research and Development Hq USAF initiated a comprehensive study of aircraft technology, under the leadership of Col. Bernard Schriever, who was assisted by Gene Root, head of the Aircraft Division of Rand. Among the conclusions of the study was, if intercontinental bomber operations was desirable, and if high performance in the target area was necessary, air refueling was inevitable. Results were published in a document, "Development Planning Objectives (DPO) for Strategic Bombers." When this study was briefed to General LeMay by Colonel Schriever and Gene Root, he reacted strongly. Consequently, when it was announced that Hq USAF was initiating such a DPO study of strategic missiles, General LeMay ordered his staff to attend study sessions so we could get SAC's views injected into the study process at an early stage and thus avoid rude surprises when the study was completed.

Colonel Nielsen asked me to name an operations analyst to go with him to the first session, and I chose Ed to participate. This was a unique opportunity to become acquainted with the state of the art in all phases of missile technology. A colonel was the USAF head, and Dr. R. W. Krueger, head of the Missiles Division at Rand, was his technical assistant.

Bob Krueger and Ed Williams of Rand would write up their notes each day. Procedures were to invite leaders of industry and government laboratories to brief their assessment of the technical approach and preliminary design of different types of missiles, for example aerodynamic, ballistic, air-to-surface, surface-to-surface, powered by a variety of propulsion systems such as ramjets, turbojets, liquid and solid rockets, and nuclear rockets. Navigation systems were of keen interest, particularly the potential of inertial navigators. Warhead technology was of great importance to the ICBM technology, as weight limits were critical.

Sessions were held most of the week on alternate weeks, so Ed spend almost every second week in the Pentagon for most to the winter of 1953. After each week of sessions he would come back to Hq SAC and brief appropriate individuals in Plans, Requirements, Materiel, and sound out his associates in Operations Analysis so he could take back responses to Bob Krueger and Ed Williams, which they would include in their notes and inputs to the study.

In particular he took a curve of weapon yield versus accuracy, which we would consider marginally acceptable for an ICBM warhead. The curve was developed from a background analysis by Rod Smith and Paul Hower. Ed would brief inputs he was taking back to Operations Analysis staff. He recalls, "Such feedback was an invaluable sounding board. We would get Zim's and Stan Lawwill's reaction and their knowledge of the attitude and opinions of General LeMay and the senior staff. Then the inputs would be briefed to Operations Plans, Requirements, Intelligence, Plans, and Materiel before they were taken back to the Pentagon."

Because of the various strategic missile development programs such as SNARK, NAVAJO, and ATLAS there was much coordination to be done with SAC. Ed became a member of an informal missiles team, consisting of Lt. Col. John D. Bartlett of Plans, Lt. Col. Dan Maloney of Requirements, and Maj. John W. Barter of Materiel. They attended many briefings at contractors' facilities and at R&D locations, regarding training, maintenance, supply, basing, and operational concepts. Later Lt. Col. Gary Fremouw of Air Installations became deeply involved when serious planning about base facilities began. Operations Analysis worked closely with him and his people.

We in Operations Analysis were particularly bothered by USAF plans for a soft basing concept which would be facing possible attacks by large yield thermonuclear weapons. The two seemed incompatible. Ed began a campaign to recommend a basing concept which would be a combination of hardening and dispersal, based on Operations Analysis studies. That informal missile group concluded that locating a wing of strategic missiles such as SNARK at Limestone AFB, now Loring AFB, in Maine, which was already a major bomber base and weapon stockpile site, was not sensible. Plans were to stock the base with 90 or 100 missiles, with only a handful to be assembled and ready. The rest would be stored in eight or nine original shipping containers for each missile. Obviously they could not be assembled, or evacuated, before or under attack on short notice.

Lieutenant Colonel Fremouw and Ed talked about this situation for several days before the senior staff Facilities Review Board was to meet. Colonel Fremouw asked Ed to prepare a briefing and give it to the board, headed by Maj. Gen. Archie Old, director of Operations. With help from others in Operations Analysis the briefing was completed about 2 a.m. on the day of the briefing. In General Old's absense, Maj. Gen. J. P. McConnell was chairman.

The briefing recommended that the wing be split into 10 groups of missiles, based in 10 different locations off base around the countryside in a "ready" condition. Each launch facility should be spaced and oriented so that a single incoming weapon could damage no more than one launch facility. After the discussion after the briefing, General McConnell said, "Let's go for it."

This started a large effort to change the basing concepts for strategic missiles. The SAC informal missiles team took the briefing to Hq USAF, R&D; to the Air Material Command Council, headed by General Chidlaw; to Patrick AFB, Florida; to all the major aerospace contractors involved; and to the Ralph M. Parsons Company, who were the architectural engineers selected by Hq USAF to develop plans and design the first strategic missile bases. The team was asked to present the briefing to the Air Force staff in the Pentagon.

The briefing was given to General LeMay first. Ed looked forward to that event with some trepidation, as it would have been quite embarrassing if he were to disagree with the position after all the presentations around the country. The briefing went smoothly. General LeMay puffed on his cigar, sat quietly for a bit, and then asked, "How are we going to train with these birds?" The answer was by simulator as much as possible, and by expending some missiles. It was a question which is still difficult to answer to the satisfaction of top commanders.

They pressed on with their planning. The general concept of hardening and dispersal has prevailed to this day in our strategic missiles programs, so Operations Analysis made a contribution to the national military posture. We had the confidence of leaders like General LeMay who permitted us to do our reflective thinking and professional analysis, and were willing to listen to and seriously consider our conclusions and recommendations.

Richard J. Camp

Rick Camp was at Hq SAC, Omaha, from 1952 to 1955 and then again from late 1959 to 1964. I first met and interviewed Rick at the University of Iowa in 1951. Early in 1952 he agreed to join me at SAC.

During his first year he was primarily involved in developing attrition models under the guidance of Paul Hower. When he looks back on those efforts now, he realizes the early models were crude (though not for the state of the art at that time). However, they certainly helped in our later development at Hq SAC of some rather large-scale, sophisticated simulation and probabilistic models. The work also contributed to subsequent models, such as the Penex Model, and the simulation war games which were used for at least 20 years to study and evaluate proposed war plans.

In those early days one of the problems that we faced was that of providing to ourselves and to new operations analysts training in specialized techniques and capabilities applicable to the analysis of air warfare. Much of our time was spent in the development and explanation of models for use in such things as determining bombing accuracy, damage from nuclear weapons, probabilities of kill, confidence intervals, and other factors.

Rick's background in mathematics and statistics proved more useful than he had dared hope in addressing interesting problems in air warfare. In those days the understanding of statistics was at a much lower level than today among the military personnel with whom we worked. It has been of considerable interest to observe over the years, since our first days in Operations Analysis, the use of the techniques of operations analysis in providing illumination and inputs for decision-making. Their use by military personnel has widened considerably. It is commonplace now to see military officers using not only basic probability and statistics but optimization techniques, linear programming, dynamic programming, inventory theory, game theory.

Attrition studies led Rick naturally to contacts, associations, and friendships with many of the stalwarts in those days in operations plans, tactics, and current operations — men such as Bill Crumm (later Major General Crumm), Eric McDonald, John Fahringer, and many others. When John Fahringer asked us to help him in the analysis of the B-47, our newest bomber, against the MIG-15, Rick jumped at the opportunity to be involved in this historic analysis. Our first MIG-15 was turned over to us by a Polish pilot.

In those days many facilities at the Air Proving Ground Command, Eglin Air Force Base, Florida, were devoted to the collection of data on operational tests. Both an Air Proving Ground Command pilot and a SAC pilot put the B-47 through all sorts of maneuvers to evade the MIG-15. Rick says he was fortunate to accompany most of those tests with his stop watch, and without his parachute on, since the area behind the co-pilot's seat in the B-47 was very tight. He watched over and over again the failure of the B-47 to evade the MIG-15. Although the MIG-15 pilots, several of whom were Korean aces, were not happy with the comfort or the controls of the aircraft, they all gave high praise to its maneuverability and its capability to stay with and outmaneuver the B-47.

Rick will not soon forget one mission when the B-47, with the SAC pilot at the controls, did indeed find its limits in maneuverability. Flying at about 42,000 feet the pilot continually increased the bank angle, turning left and right, in an attempt to evade. Rick had been briefed earlier that in the case of an emergency, his procedure was to wait until the pilot and the co-pilot had ejected, and then he was to go out through the opening.

He believed it necessary when they went into a spin at 42,000 feet to try to get his parachute over his shoulders. He got the chute over one arm, but, as he had heard earlier from Fred Fennema, there is such a thing as "g" forces involved in a spin. During the next two or three minutes, before the pilot regained control, Rick found he was unable to get the chute over his other shoulder. Somewhere along the way he dropped his watch. They arrived safely back on the ground. The MIG pilot during the debriefing wondered what in the world they were doing.

It's fair to say that the outcome of the analysis proved to the satisfaction of all that the B-47 could not outmaneuver and evade the MIG-15. Other types of bomber defense techniques would be required.

Later on, during Rick's three years at Hq Fifteenth Air Force, March Air Force Base, he was asked to return to Eglin to head the analysis of the B-52 against the MIG-17. This time there was no suggestion that the B-52 could outmaneuver the MIG. These tests were of the electronic countermeasures equipment, including chaff, which had been developed for the B-52. Earlier the tentative conclusion had been reached that the B-52 could survive the MIG-17 because of its capability to carry large quantities of bomber defense equipment. The conclusion was borne out by the test. Later it was proved in Vietnam.

The foregoing reminds me of a study done earlier by Rick, which General Power found useful and carried with him for a number of months. During the earlier days there were many discussions as to whether the B-52 could live in a sophisticated defense environment. In early 1953 Rick submitted a study to me, subject "B-52 Liveability." He showed that because of the capability to carry large quantities of bomber defense equipment, the B-52 would be able to live. Though some of the models were crude in retrospect, the study did hit some of the important points with respect to B-52 liveability, and was well to the point, as the B-52 is still with us.

Another example of one of the early contributions of Operations Analysis to SAC was the use of game theory in an early missile study. As the MINUTEMAN started to become operational, there was considerable concern about the possibility of the so-called "pin-down" tactic. A Soviet incoming attack might be so spaced as to prevent the launching of the missile, thus permitting Soviet bombers, with great accuracy and warhead yield, to come in and destroy the missiles in their hardened sites.

It was considered necessary to think about the possibility of firing out during a "fire-in" Soviet missile attack. Rick was chief of the Tactics and Planning Division. The request from Operations Plans for assistance had, to Rick, all the elements of game theory. There were several different "fire-out" and several different "fire-in" strategies. If one sets those up as a game matrix, one sees that there is a "saddle point" and there is an optimum strategy for both the enemy force and for ourselves. Gerald Berndt was the chief analyst on this study.

The results of the study did suggest a clear optimum "fire-out" rate for our forces, and they were used for years as the basis for the "fire-out" policy of the missile. It demonstrated a practical use for the abstract concept of game theory. Game theory was originally developed to evaluate economic behavior, but has been used much more extensively in strategic exchange and arms control studies.

Frederick F. Fennema

Fred Fennema was an operations analyst with the Twentieth Air Force in the Marianas in World War II and immediately after the war, with the Alaskan Air Command. He joined SAC after the headquarters moved to Omaha. Here are some of his comments on his activities:

> There is nothing so powerful and rare as perception of several previously independent ideas or circumstances which can be converged to yield a useful synergistic result. Now that I look back on General LeMay's performance, I think the above was his greatest talent. I think it was this quality which made him appreciate Operations Analysis products.
>
> The B-47 low level penetration and toss bombing project is a case in point. [This was suggested and fostered by Fred.] We recognized the problems posed by the "missile gap" — and the Communist Manifesto which justified any unprovoked means, including total missile assault, to achieve world dominance. We also realized the drain on the USSR budget to achieve a superior strategic posture — but! — their nakedness in terms of low level defense capability. Ergo the suggestion of low level, tanker-supported B-47 operations from African (and other) bases.
>
> Develop the capability in secret? No. Demonstrate it? Yes. Why? We wanted the Soviets to have to devote a highly disproportionate part of their budget to building a low level defense capability — more importantly, we needed to gain time until the missile gap was closed.
>
> I briefed on 100-foot altitude penetration, Initial Point, and target "break-up." I had asked for and received a series of pictures of a B-47 radar scope on sector scan on a run, approach, and landing at MacDill AFB. From these, with extrapolation, we determined the feasibility within the 2.0 to 2.6 "g" limit load factor (envelope). The existing equipment would give an immediate capability, with better or modified equipment as necessity demanded.
>
> An Eglin AFB Task Force was formed, given high priority, and you know the story from there — "those crazy chimney and hilltop flying airplanes." Low level testing of B-52s was also begun for penetration only.

There were other projects with which I was involved. B-29 losses to MIG fighters in the Korean conflict were unacceptable to Hq Far Eastern Air Force (FEAF). You had previously sent at least two analysts on 3-month rotations to General Bill Fisher's FEAF Bomber Command. When I arrived in Yokota I had the MIG/B-29 loss problem dumped on my desk. I started with the Hq FEAF and FEAF Bomber Command tactical doctrines, and the intriguing and frightening phenomenon of MIG non-firing passes through tight B-29 formations flying Shoran arcs. An analysis was made. Wing commanders were briefed.

General Irvine (Carswell) requested my running what was called the B-36 operational engineering program. The first step was to design a systematic record-keeping system for collecting and analyzing all failures by type and frequency. Another was a combat profile sortie-rating system, which in about two months went from a success factor of 6% to 78%.

There was the "weather" balloon armada with a photographic package to travel with prevailing winds across the USSR at automatically controlled variable profiles of 40 to 60 thousand feet. A nominal seven-day trip was calculated with the photo packages to be picked up by parachute-snatching C-82s over the China Sea and environs [Project GENETRIX].

Project GENETRIX

The mission of GENETRIX was to obtain photographic and electronic reconnaissance of the USSR and its satellites using Weapon System 119L. Weapon System 119L was developed by Air Force agencies and civilian institutions for the purpose of conducting pioneer reconnaisance by the use of high-altitude balloons. The basic concept of operations was to launch from Western Europe balloons which were expected to transit the target area in seven to ten days, and then be tracked and recovered within the Far East and Alaskan areas. SAC was assigned this responsibility in March 1955. Then Brig. Gen. William P. Fisher was chosen by General LeMay to direct the project, as commander of the newly formed First Air Division. Operations Analysis personnel participated.

The project was remarkably successful. A total of 13,813 usable exposures were obtained, covering 1,388,745 square miles of the Sino-Soviet area. Major General Fisher (USAF Ret) wrote to me in March 1981:

> Operational control for the First Air Division was established at Offutt AFB. Launching sites for the balloons were set up at various locations in Europe, and recovery squadrons in the Japan-Okinawa area. All finally was arranged and operational and we began launching.
>
> General LeMay visited our control room several times a week, made suggestions and comments, and in general gave us moral support. He gave the First Air Division top priority, loaned us Operations Analysis assistance, his staff backup, and all the field assistance we needed. This photo and electronic reconnaissance effort, soon followed by the U-2, produced much valuable and original information.
>
> For a positive organization like SAC and a super-positive commander like General LeMay to undertake such a non-positive, uncontrollable, and questionable method of obtaining information as GENETRIX exemplifies his dedication, loyalty, and willingness to try even the most tenuous methods of getting a needed job done.

Besides Operations Analysis assistance to the First Air Division at Offutt AFB, assistance was provided in the field. Initial early training launches from Loring AFB resulted in a high failure rate. General Fisher asked for assistance from Operations Analysis. Fred Fennema wrote to me in February 1981:

> General Fisher asked Operations Analysis assistance to establish cause (failure mechanism), find a fix, or predict the success of the units scheduled for launching from Western Europe. General Fisher remembered my assistance while he was Commander FEAF Bomber Command during the Korean war. He had suggested to General LeMay that he arrange with Commander USAFE, Ramstein, Germany, for my indefinite loan to work on the problems there.
>
> Through analysis of aerial photos of bursting balloons I established the burning mechanism. I statistically determined what the overall success rate would be — within 3 of the actual.
>
> I was also in the Command Post in England when the full-scale operations began.

Advisory Groups

For decades it was quite the vogue for governmental agencies to have "outside" advisory groups on myriad subjects. They were "outside" in that they were composed of experts not assigned to that agency. The Department of Defense, particularly, had many expert advisory groups. The membership was almost strictly civilian. Most advisory groups were ad hoc. A few were "permanent."

I'm reminded of my favorite definition of expert: "X" is unknown. A spurt is a drip under pressure. Therefore an expert is an unknown drip under pressure. Experts like to use statistics. My favorite definition of statistics: statistics are like bikini bathing suits. What they reveal is interesting, but what they hide is critical.

In the definition motif: Differences of opinion equal:

In Congress:	Congressional debate
In the Supreme Court:	Judicial deliberation
In the university:	Academic freedom
Among the military services:	Petty bickering

Communication, particularly among diverse groups, is sometimes most difficult. The same spoken

Meeting in Washington, D.C., 1950s. Front row L to R: XXX, XXX, Tom Chappelle, Stan Lawwill, Percy Carr, XXX, Zimmerman. Second row at far left: Dan Leivick. Behind Dan: Bill Greene.

or written words can, and do, mean different things to different people. I shall illustrate: A young, beautiful, vivacious, and voluptuous heiress decided she wanted her full figure painted, in the nude, by a famous artist. She suggested this to him and offered a substantial fee. He was horrified — it was against his morals, ethics, and demeanor. She doubled the fee. He pondered, and said he would think about it over night, and phone her in the morning. In the morning he phoned and accepted. When she appeared at his studio at the appointed time, he asked, ''May I keep my socks on? The floor is cold.''

Once Viscount General Montgomery drew on a blackboard for me this matrix:

	LAZY	INDUSTRIOUS
BRILLIANT	Staff Officer	Field Commander
STUPID	Menial jobs	Look out!

The brilliant industrious officer was a good field commander; the brilliant lazy were good staff officers; for the stupid lazy one could always find menial jobs. But look out for the stupid industrious ones!

Among my own observations concerning ''experts'' are the following:

> Scientific Method — One fact can upset a hell of a lot of theory.
>
> Platitudes — Labeling platitudes Secret does not make them any more noteworthy or worthy.
>
> Airports — If you like to hike, fly.
>
> Revealing information — If one bares one's bosom often enough, one is certain to end up getting screwed.
>
> Expertise — Experience without expertise is meaningless.

* * *

In some instances it is difficult for ''insiders'' to influence some other ''insiders.'' Shortly after World War II I generated a proposal, pointing out that the

OA Technical Conference, Florida, 1955. Seated to Zim's right: Max Oldham. Seated to Zim's left: Roy Brothers.

greatest problem of B-29s was lack of range. Nature had not been considerate enough to put islands everywhere needed for ground refueling. Even if she had, political considerations could deny them to us. This has been subsequently demonstrated many times. I suggested that in-flight refueling be investigated. I gave some preliminary calculations showing payoffs. The endorsement back, signed by a major general, suggested I stick to my scientific knitting, and the military would take care of operational problems.

Had the general and I, at that time, had the slightest

Some members of Scientific Advisory Committee to Secretary of Defense, 1956.

inkling of my subsequent influence on operational considerations, and operational concepts, it would have scared the hell out of the both of us.

Hap Arnold asked a group of ad hoc experts to assess the military potential of the turbo-jet engine. Their report said: No way, much too heavy, no range, etc.

I was selected to be a member of the Killian Committee for six months in 1954-1955. It was then a Top Secret Presidential Committee. Our mission was to recommend to the President how science and technology could reduce the effects of a surprise attack on the United States. We worked in the Executive Offices of the President. Dr. Killian was on leave from his position as chancellor of MIT.

Within a few days after I reported the Navy arranged for Commander Pete Aurand to be assigned to my group. He was very sharp. We developed a mutual respect. We also became spirited antagonists on some issues. I learned beyond doubt that each day the Chief of Naval Operations was in town Pete would either have lunch with him in his office, or phone him.

Whenever Pete came into my office the first thing he would do was turn the water hemisphere on my globe of the world toward me. The surface of the world is only 29 percent land. After he did this a few times I said, "A hell of a lot of the earth's surface is covered by water, isn't it?" After a slight pause, "I wonder how much of the earth's surface is covered by air?"

Pete was among the first to fly faster than sound. He always wore civvies, never a uniform, during his duty with the Killian committee. When visiting Hq SAC, with all the other military members of the committee in uniform, General LeMay asked Pete, "Where is your uniform?" Pete pulled a copy of a special order out of

With other members of the Scientific Advisory Committee to the Secretary of Defense, July 1956. From left: Maj. Gen. Bernard Schriever, Si Ramo, Maj. Gen. Kern, Zimmerman, President of Rand Corp.

The Scientific Advisory Committee to the Secretary of Defense, visiting General Dynamics, 1950s.

Visit to North American Aviation prior to its becoming Rockwell. Dutch Kindelberger, whose team designed the P-51 Mustang in 1940, is on the extreme left.

his pocket, signed by the Secretary of the Navy, authorizing him to wear civvies during his tour with the committee. This was prior to the option later adopted that an officer could wear a uniform or civvies, except for one specific day a week, when he must wear a uniform. That option was for Washington headquarters types.

On his civilian jacket lapel Pete wore a device which said "MACH BUSTER." He abandoned it when I started calling him Buster.

One day I told Pete that I was pleased that his son was going to be in the Air Force. Somewhat startled he asked, "Why do you say that?" My reply, "Your father was in the Army, and you are in the Navy. It appears that each generation is getting smarter."

Soon after the committee work was finished, Pete was promoted to captain and assigned as Naval aide to President Eisenhower. Like his father, Pete retired with three stars.

I lived in Omaha while with the Killian Committee, and I worked full time for over six months in Washington. I had to maintain my family in Omaha along with the added expense of my living apart from them. Besides his salary, paid by his agency, each member received either actual expenses or a per diem of $15, which his agency also paid. I was the only civil-servant member, and I received a per diem of $9. Even at that time, living and eating in Washington on $9 a day was impossible. And we took trips. I could not stay around the corner in a fleabag when my colleagues were together in a first-rate hotel.

I kept complete records of my expenses and reported them honestly and deservedly on my income-tax return. My return was questioned by the Omaha Internal Revenue Service office and I was dunned an additional sum, before any clarification, or justification, of my figures was requested. Visiting the IRS office I found that handling my return was an old GS-5 woman. She as much as told me that she had a GS-17 in her hands (I had not yet been promoted to GS-18), and she was going "to give it to me." She told me she knew that "her boys" made money when they traveled. After much haggling, I succumbed to a figure one-half the extra amount originally dunned, even though it was unfair to me, and at that time was quite a financial strain.

* * *

Dr. John von Neuman was chairman of the Air Force TEAPOT Committee. At his invitation I became a member. The deliberations of this committee resulted in the establishment of the Ramo-Wooldridge Company, and the development of the THOR and ATLAS missiles.

Shortly after TEAPOT von Neuman was asked to form and chair a Scientific Advisory Committee to the Secretary of Defense. Again, at his invitation I became a member of the committee. During the entire history of the committee I was the only one directly associated with a military service. The others were from governmental agencies, industry, laboratories, and the universities.

Besides THOR and ATLAS, the committee advised the Secretary of Defense on other missiles of all three services: NAVAJO, SNARK, RASCAL, HOUND-DOG, GOOSE, JUPITER (Army), and POLARIS (Navy). In 1959, when I left SAC to report to Hq USAF, I submitted my resignation to the committee. This was custom, as perhaps the Secretary of Defense would not want someone in a new position to remain on the committee. My resignation was not accepted. This caused me to believe that I had succeeded in being objective when dealing with Army and Navy, as well as Air Force, problems. I served eight years, until Mr. McNamara abolished the committee in his first year as Secretary of Defense.

I received a citation from the Secretary of the Navy for my work on POLARIS while I was being paid by the Air Force. One of my contributions was to discover early in the program that the Navy was allocating part of the development cost of the POLARIS to the sub-hunter-killer role. This would make its cost look better in the strategic-bombardment role. I pointed out to the Navy that they would be the first to scream if someone suggested the hunter-killer role for POLARIS. They adjusted their figures.

My citation served another purpose. I put it on the wall of my office when I went to the aerospace industry. When an over-zealous engineering or management type would come to me to ask that I prove his preconceived notions, I would point to the citation, and inform him that I received that while being paid by the Air Force.

National Security

National security and the protection of classified information are personal responsibilities. Two requirements obtain for giving classified information in any form to anyone. The individual giving the information must have proof that the recipient has valid security clearance at the proper level. The giver is responsible to ascertain that the recipient has a need to know. Proper receipting, proper care and storage, proper packaging, proper transmission — all are a part of each individual's responsibility.

Air Force Regulations 205-1 and 205-15 were the primary documents spelling out the responsibilities for security and the handling of classified information. Each member of SAC, military and civilian, was required once a year to take a written test on these and other pertinent regulations. The only passing grade was 100. That was a SAC requirement, not required throughout the Air Force.

One day I received a call at my desk at Hq SAC at Offutt AFB from an employee of an Air Force contractor. He was not a friend or a colleague, but I knew the company for which he worked, and the classified SAC project on which he worked. Stan Lawwill had been in meetings with him for which a Secret clearance was required.

The person said he had an unexpected layover for several hours at the Omaha airport. This was pre-jet. Could he come out and see me? I replied, "I do not have a security clearance message for a visit. We could talk only unclassified. I'm rather busy. Perhaps some other time." Stan Lawwill was not available.

He did not come. It was Morton Sobell, who got 30 years when the Rosenbergs were executed. That illustrates that it could happen to you. The FBI found in his notebook "Call Lawwill." They did not know who Lawwill was, but they soon found out.

We were so pleased that we had been properly schooled on security. Not only because we avoided personal embarrassment and trouble, but because we were not a party to giving those bastards any information. Morton Sobell was released from prison several years ago.

* * *

Unfortunately, some have over-reacted in security matters. We all remember the witch hunt, the McCarthy hearings. We are either old enough to have seen him on TV and read about him at the time, or young enough to have read about him since.

A very competent mathematician had been with me early on the beachhead in June 1944 and within several miles of the front line across France, Belgium, and Germany. When one lives with a man for over a year in danger in the combat zone, twenty-four hours a day, seven days a week, there is no artificiality, intrigue, suspicion, or uncertainty in the relationship. I would have bet my life on his loyalty, and once almost did.

I persuaded him to join my staff at Hq Continental Air Forces immediately after the war. When I joined Hq SAC on the day it was established, he joined me. This gentleman had a Top Secret clearance. At that time a "Q" clearance was needed in addition for access to nuclear information. I had him apply for a "Q" clearance. It was denied.

I probably wasn't supposed to, but I acquired a copy of the clearance denial. He had been on the faculty of a California university. One year a summer hire permitted a mimeograph machine to be used to reproduce some Harry Bridges propaganda. My man did not know about it, but because he was in charge he was held responsible. That resulted in the denial of the clearance. Also in the report was a statement that he was not a very good teacher. What relevance does that have to loyalty?

He could have stayed with his Top Secret clearance, without access to nuclear information, because his Top Secret clearance was not rescinded. However, his loyalty to me was such that he resigned, because he did not want in any way to cause even slight embarrassment or damage to me.

That action ruined his life. The stigma, though not valid in my opinion, stayed with him. He had difficulty getting a position to utilize his talents properly. This led to his divorce. Subsequently he completely disappeared from my view. A witch hunt prior to McCarthy.

* * *

Early in SAC when General LeMay visited a combat unit he found the one guard at the flight line eating a ham sandwich. He was quoted, correctly, as having said that a ham sandwich was guarding the aircraft.

In 1951 he instituted a special course in SAC security. It emphasized protection of bombers on the ground and protection of key personnel. Picked penetration teams were trained to effect "sabotage." "Enemy agents" posed as SAC officers, as enlisted men working on the flight line, as vendors who serviced soft drink machines, fake mailmen, phony air policemen, phony contractors — SAC penetrators have worn every disguise. It was a very dangerous job, and the people trained with dedication and devotion.

There is a story that some telephone repairmen came into General LeMay's office. He became suspicious and held the men against the wall with an automatic until the Air Police arrived. General LeMay does not remember that happening. Perhaps the story stemmed from a true incident when a major knocked a phony telephone repairman to the floor, and held an automatic against his temple until the Air Police arrived.

Besides these SAC-trained penetrators, Air Force teams from outside SAC would try to penetrate SAC bases. Col. Mike McCoy was an outstanding combat commander and leader of men. I was visiting him one time, when he commanded a combat wing at Orlando AFB, Florida. When an Air Force team from outside SAC was going to try a penetration, they talked prior to the action to the commander, in order to reduce the chance for accidental casualties. It was with the commander only, and he was pledged to secrecy.

The penetration of Orlando AFB was planned for the next morning. Mike convinced the outside briefers that I was trustworthy, and invited me to sit in on the meeting with them the night before the planned penetration.

Mike was magnificent. He said, "I wish you luck. I hope you don't run into trouble in the scrub outside the perimeter fence. You know this county has the greatest density of rattlesnakes in the world. Occasionally some alligators roam around." He continued, "I think my young air police can handle their new sentry dogs." And so on.

I often wonder how much heart the penetrators put into their operation.

Unfortunately, later Mike was killed in an aircraft accident. Orlando AFB was renamed McCoy AFB.

Dayton, Ohio, Bombing Exercise

Shortly after General LeMay assumed command of SAC, in October 1948, he asked for a briefing on the command's bombing capability and accuracy. I prepared the briefing, with particular help from General J. B. Montgomery, Director of Operations, and briefed General LeMay. The capability looked good from the information available to me. I had not been responsible for generating the data, but only for assessing it. It showed the radar bombing results with a Circular Error Probable (CEP) of 1,500 feet. This meant that 50 percent of the bombs would be expected to fall within 1,500 feet of the target.

General LeMay did not believe the results. He ordered J. B. Montgomery (Monty) to lay on a practice attack on Dayton, Ohio. Third and Main Street was chosen as the primary target. He chose Dayton because Wright-Patterson AFB was there. We helped with the planning.

A Maximum Effort was conducted against Dayton. The weather, admittedly, was not good. But not one aircraft finished the mission as briefed!

Some things needed to be done, and they were. When General LeMay left SAC in 1957 every city in the United States of 25,000 or more population had been bombed frequently. San Francisco was bombed over 600 times a month. The bombing results were quite satisfactory. Of course, the CEP was determined by radar bomb scoring equipment on the ground, and no bombs were dropped.

Col. Walter Coons had noted the lack of ability of SAC to hit the NCR building in Dayton. He reports:

> As a result, Clarence Irvine was brought up to the Headquarters and General LeMay asked him why we were having so much trouble. At that time I popped off and said, "It seems to be a good target, and we seem to see it all right on radar, but why we don't hit it, I don't know." To which Irvine said, "If you are so damn smart, you come down and fly with me and see what you can do." So naturally all I could do was to get out my old flying suit. We took off that night for the 509th.
>
> The next day with P. K. Carlton as pilot we took a B-50 and made our run on Dayton. We had an Initial Point (IP) about 50 miles west of Dayton. I figured this was either sink or swim. Using the old Q-23, which was in the B-50s in those days, I made an offset from Wright-Patterson AFB, which I knew we could see very well on the radar scope. We came off the IP with about a 150-knot tailwind, which put us across the target at a very respectable speed. About halfway down, I did lose the NCR building, so I threw the Q-23 into offset and made my run on Wright-Pat. We got a very fine bomb, something like 500 feet, on the Radar Bomb Scoring (RBS) run, which in those days was outstanding.

Stan Lawwill, too, wrote to me about the Dayton exercise:

> I remember spending several nights in the RBS plotting van, after the initial Dayton exercise — once with Bill Irvine and once with Jack Ryan when he was CO of the 509th. Bill was insistent that his B-36s [the 7th Bombardment Wing, Heavy] would demonstrate the best bombing that had ever been scored by RBS. When the first B-36 checked in over "orbit point able," we switched the plotting board to expanded range and recorded the B-36 position. When the bomb run reached the max scoring distance (expanded scale) we switched back and recorded the end of the bomb run and "bombs away" (simulated). Under Bill's close scrutiny I scored the run; it was pretty good.
>
> When the next B-36 checked in over "orbit point able" we repeated the entire process. That B-36 was recorded to be many miles away from the first's "orbit point able," its ground track was remarkably different from that of the first B-36, and it, too, scored well. By the time the night was over that plotting sheet had tracks all over it from 11 B-36s that were supposed to be flying the same mission — same Initial Point, same altitude, same target. Fortunately, there was almost no wind shift that night, so all the differences were due to crew differences. Although their bombing scores were good (not excellent), the crews had a mission debriefing from Bill which stressed discipline — "Do exactly what I tell you; approximations are not acceptable!"
>
> When the 509th flew against Dayton, Jack Ryan was there in the plotting van with me all night. He had his notebook and personal computer (actually a circular sliderule). He not only checked everything from the Rawin runs (several that night because the winds were shifting), to yaw angle, ground speed, and bombing error of each aircraft. He insisted that he was there to learn — for himself and his crews. He made careful notes on each bomb run through the whole mission (20 or so aircraft, as I recall). After breakfast, I went to bed. Jack climbed into his aircraft and returned to Ft. Worth.

* * *

SAC's mission in 1959: Be prepared to conduct strategic air operations on a global basis so that in the event of sudden aggression SAC could immediately mount simultaneous nuclear attacks designed to destroy the vital element of the aggressor's war-making capacity to the extent that he would no longer have the will nor the ability to wage war.

* * *

Col. Lawrence J. Israel (USAF Ret) has an additional view of the Dayton exercise. He wrote:

> Perhaps the very noticeable first event which occurred some months after General LeMay's assumption of command was what was called the Dayton Bombing Exercise. It was directed by General LeMay to test the capability of the command. Dayton was picked as the site — for one thing it has many surrounding towns and for that reason should present a non-complicated, or simple problem in radar bombing, as opposed to a point target which sometimes is difficult to identify.
>
> The Dayton complex should have provided a rather easy target because of the many references. Whether or not this was an actual consideration I do not know. It brought out a serious deficiency in the bomber crew training. With a suitable ground map the presentation on the radar bombing scope should have been rather easy to identify, providing a proper amount of study had been undertaken by the bombing crews before the bomb run. It provided an ideal situation, but as it turned out it was a horrible fiasco. The scoring was

accomplished by Radar Bomb Scoring (RBS) radar on the ground.

The main difficulty with the Dayton Bombing Exercise was the tendency of the crews to mistake some of the towns around with Dayton itself. Many of the bomb runs were made on towns other than Dayton. The CEP for this bombing exercise was, as I remember it, 12,000 feet. This was horrible. The capability of the bombing systems was much more accurate than that.

General LeMay's reaction to this, as one might expect, was rather violent, or certainly vehement. Events began to happen so that in two years, as I recall, the CEP had been reduced from 12,000 feet to 1,200 feet. This was primarily due to the very strict training program which was imposed by General LeMay. It is to General LeMay's great credit — perhaps I should point out that he was the greatest man I have ever known — and having been there at the time, that the CEP of 12,000 feet was reduced by such an order of magnitude.

It is a matter of fact and perhaps of interest, that a friend of mine, a bombadier-navigator, reran that same route with much pre-study, and his CEP was 500 feet. The bombadier was Walter Coons.

Bootstrap Operations

One outstanding attribute of early SAC was its inclination and ability to solve problems within its own resources. These problems were often imposed on SAC. Often they were the responsibility of another Air Force agency. There are many examples throughout this book. There are others that are not covered, for example "The Blue Sky Realty Company."

Two examples are given by Col. Larry Israel:

As large a job as the improvement of crew CEP was, it might be called a small problem compared to the bombing equipment status. There were many reasons for that status. Having been involved in improving the equipment, I can say that one of General LeMay's important contributions to SAC was the improvement of the capability of the equipment. By capability, here, I don't mean accuracy, but I mean its ability to operate over the period of one mission without failure.

At that time something like 50 percent of the aircraft that took off on a training mission returned because of malfunctions in the radar bombing equipment. The cost just to take a B-36 off the ground was high. Thus aborts were quite costly. Also aircraft were on the ground which couldn't take off because of malfunctions in the bombing equipment. Lack of spare parts contributed to the problem.

General LeMay got personally involved and instituted a very large Air Force-wide attempt to correct the situation. The culmination of his actions resulted in a meeting at Hq SAC on 2 February 1951. The meeting was to brief then Maj. Gen. Donald Putt, who was then Director of Research and Development, Hq USAF, in the Pentagon. Bell Laboratory personnel, including Don Quarles and four or five of his top people, were present. [Authors note: I also was present.]

The briefing was given by me. Upon completion of the briefing, and some comments, General LeMay took the floor and read the riot act. He had plenty of reason to do so. Also in the meeting was Col. Gordon Blake, who was chief of the Air Force Armament Laboratory at Wright-Pat AFB. He was ARDC's [Air Research and Development Command] responsible person. General LeMay gave Mr. Quarles a very hard time. This has a humorous note because later on, Mr. Quarles became General LeMay's boss as Secretary of the Air Force.

(Honor in one's own household: When Don Quarles had retired as Secretary of the Air Force he told me how much he admired and respected General LeMay. He said his, don's standing in his own household was considerably enhanced when he received a warm Christmas note from General LeMay. His family considered this the most important item he received that Christmas.)

Colonel Israel continued:

General LeMay said to Mr. Quarles that if Western Electric made washing machines the way they made bombing systems they wouldn't stay in business for two months. Mr. Quarles responded in a manner. Amusingly, about three weeks later there appeared in the *Saturday Evening Post* a full page ad by Western Electric which said, "No, we do not make washing machines but we do make other things." Following was a brief outline of what Western Electric did make.

This meeting created a considerable amount of activity which continued for several years. It was a very large scale program. It resulted in two projects which went by the names Wesco and Sperco, because both Western Electric and Sperry were involved in the radar bombing business at that time. To their credit Western Electric exerted every possible effort, at several levels, depending on the nature of the problem at hand. The minimum time for the rehabilitation of the bombing equipment was 37 days. General LeMay was very sensitive to aircraft out of service.

The complexity of the systems can be indicated by the Q-24 having something like 450 vacuum tubes. The Sperry system, though it was "mechanical," had 75 more, and weighed considerably more. They were, and this should be emphasized, very complex pieces of equipment. In those days anything with 450 vacuum tubes was complex. The vacuum tubes were only a small part of it; there were many other components involved in the entire system. Sophistication seemed to be the mode, ignoring the complexity and the resultant unreliability.

The Q-24 was a beautiful piece of equipment. Having spent three months at Bell Laboratories, eight hours a day, studying nothing but one bombing system, made me quite familiar with its intricacies and complexity. It was too complex to put into an aircraft.

Perhaps a ground installation? Yes. It was not designed for an aircraft environment.

The major portion of the difficulties resulted from failing vacuum tubes. There were several factors involved in this. I was once told by a vice-president of RCA that manufacturing vacuum tubes was more an art than a science. A finished set of tubes was divided into three groups. Those that had the highest reliability were sold to the TV and radio manufacturers. Those which reached the middle level of reliability were sold to the repair shops. Those at the lowest level of reliability were sold to the Armed Services, because they met the Signal Corps specifications at the time. That is an actual quote from an RCA vice-president to me. Thus no hearsay or secondhand information.

Before the major undertaking, we set up a test in SAC. We set up what we called an aging rack. It was a rack of sockets all connected so the tube would be subject to normal conditions. It was set on solid concrete so there was no vibration. There was no off and on. We put 50 tubes in each rack. For one particular tube, the 6AK5, 14 of 50 tubes failed in 50 hours with no stress on them at all. Another tube, the 6J6,25 out of 40 failed withing 40 hours under ideal conditions.

Stan Lawwill contributed more information about equipment failure:

Aging I know a great deal about — primarily because it was my dear friend Colonel Selbie who invented the process. I remind you that Don was Electronic Maintenance Officer of the 307th Bomb Wing, Medium, at MacDill AFB when SAC was given the anti-submarine job because we had long-range B-29s which could do sea surveillance — rather than buying Truculent Turtles for the Navy. You no doubt remember sending Jack Smeltzer and me to MacDill in August to figure out how to use the APQ5 — or the ARA5 which the Air Force had given to the Navy — to do low altitude bombing of submarines. Modifying radars and calibrating them in the noses of B-29s on the runway at MacDill in August (140°F!!) is a very good way to get acquainted. Back to aging. Colonel Selbie had been transferred to Carswell AFB as Electronic Maintenance Officer. SAC could not keep its radars operational: the mean time to failure (primarily of vacuum tubes) was unacceptably short. You remember the study Operations Analysis did.

We showed:

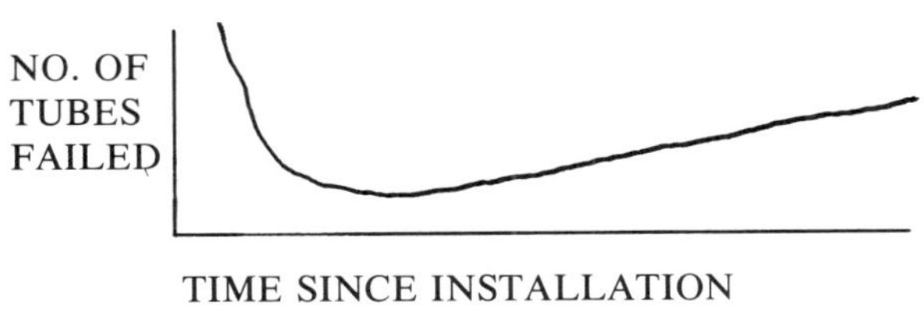

Well, it occured to Don that we ought to "select out" the early failures in vacuum tubes before they were installed in radars. So he built a big rack of vacuum tube sockets with simulated radar circuits. He installed tubes, turned on the power, and watched the failures. The tubes that did not fail in the rack were used as replacements in sets installed in aircraft. The mean time to failure increased dramatically — but still was not good enough. Don decided he wasn't stressing the tubes enough during the aging cycle, so he increased the voltages on the rack above those to be experienced in the aircraft. Many more tubes failed on the rack and those that survived, when used as maintenance parts, improved still more the mean time to failure of the radars.

I happened to be at Carswell when Don made the final modification to his tube-aging rack. He attached to the rack a vibrator that shook the aging tubes mercilessly (which, despite shock mounts, is what the tubes would experience in the aircraft). Needless to say, that process "selected out" many more failures, and SAC's radars became much more reliable.

The comptroller-type folks objected to the large numbers of new vacuum tubes that were being "destroyed" on the ground, but the operators and commanders were delighted with the increase in combat capability at such a low cost.

Larry Israel continued the story:

I went to a rather high level technical conference on vacuum tubes in New York City. I gave my briefings on the difficulty we were having with vacuum tubes. I practically begged the people there who could do something about it: Design a vacuum tube to operate in an aircraft environment. A Signal Corps general in the audience stood and read me the riot act. One question was, "How do I know these tubes will work in a Jeep?" I politely informed him that I didn't gave a damn whether they worked in a Jeep or not. My concern was for them to work in an aircraft. He gave me a lecture on his view that Germany lost World War II because they had three suppliers of vacuum tubes. The above is quoted exactly as remarks of a Signal Corps general, only to illustrate some of the conditions which existed in trying to solve this horrendous and large problem of equipment failure.

Having been in the engineering business all of my life — 21 years at Westinghouse and 18 years with the Air Force — I would be the last one to deny the engineer's right to improve his equipment to make it more functionally useful. However, during that period no attempt was made, except for a very few individuals, to try to reconcile equipment reliability vs equipment complexity.

General LeMay understood this problem, and I like to think I understood it. He gave me his unqualified support in my attempts to rectify the situation. His interest was not solely in radar. He had great interest in the armament on the aircraft. Having a personal interest in guns, which I share with him, he was quite vehement about the high failure rates of the reciprocating gun mechanism which was installed in all SAC bomber aircraft. The average firing duration was unacceptably low, something like 35 rounds. After 35 rounds the gun refused to operate, on the average. General LeMay was quite upset about this.

An amusing incident illustrates what General LeMay expected from his staff. About 1 December 1953 General Don Putt and General Leighton I. Davis, from the Pentagon, visited to brief General LeMay on some new things that might appear in SAC aircraft. The last item was the Gatling Gun, a six-barreled rotating gun, similar to the gun Dr. Gatling designed during the Civil War. The gun was being developed by General Electric under the direction of the Armament Laboratory at Wright-Pat AFB. They had such a gun set up for testing and further development.

General Putt described the gun. Its firing rate was 3,200 rounds per minute, as opposed to 147 in the guns then installed. The reliability quoted was phenomenal. The uninterrupted firing of 3,200 rounds per minute was normal. General LeMay was enthusiastic. His question was, "When are we going to get these guns in SAC aircraft?"

I recall very well General Putt saying, "Two or three years perhaps." General LeMay practically flew out of his chair. He said, "Two or three years! I can install one in my automobile hobby shop in two weeks." General Putt said, "Where shall I ship the gun?"

General LeMay turned around and pointed to me and said, "Ship it to him." General LeMay didn't give me a single word of direction after that. All I had was his "two weeks." A considerable amount of talking and traveling ensued. We decided to install it at MacDill AFB, Florida, because of weather conditions, this being winter in Nebraska.

I was very fortunate to have a new officer assigned to my division, Maj. Edward Purdy, who just the year before had graduated from the University of Nebraska with an engineering degree. I sent him to MacDill to install the gun in two weeks in a B-47. Purdy installed the gun in 10 days and it was operational. The B-47 flew and the gun fired. It took about a month more to complete the details. For instance, it was put in a new turret supplied by General Electric.

General LeMay never said anything to me, but General Power did on many occasions. His secretary sent me many recorded phone calls of General Power talking to various high level people in the Air Force. General Power told the story of General Putt saying it would take two to three years, General LeMay saying he could do it in two weeks, and SAC actually did it in 10 days.

I tell this story because it has a certain amount of humor. It also indicates how General LeMay expected his staff to act, and we did. We were able to accomplish many things because of the strong leadership of General LeMay.

In all fairness I should point out that the General Electric Company gave us considerable help at MacDill.

Requirements Committee

Shortly after General LeMay became commander of SAC, a Requirements Committee was established. I was chairman. The committee membership was primarily at the major general and brigadier general level. We were briefed on and reviewed all requirements for weapon systems, equipment, personnel, and budget.

We would approve, or disapprove, such requirements recommendations. If they involved extensive Research and Development, unusual sums of money, we would prepare our recommendations for General Power, when he was deputy commander to General LeMay. When General Power became CINCSAC he continued to decide personally such matters.

When we were briefed on programs which were based more on emotion than on analysis, the briefer would receive searching questions from me. In the beginning some resented this. In the end, most of those became our staunchest supporters.

A Requirements Division was established, but the Requirements Committee continued to function for some time after that. Col. Ariel Nielsen was the first chief of the Requirements Division. Then came Col. Jack Catton. Ariel retired as a major general and Jack as a four-star general.

Ballistic Missiles Committee

General LeMay has the reputation for being a "bomber" man, and against ballistic missiles. He is a bomber man but he was never against ballistic missiles. At my suggestion to him, long before they became operational, a Ballistic Missiles Committee was established. I was chairman. The membership was primarily at the major general and brigadier general level, but it was somewhat different from the Requirements Committee. Some officers were members of both committees.

We received information on ballistic missile research and development activities, and we kept the SAC staff informed. We considered necessary Hq SAC organization and activities to prepare for receipt of ballistic missiles in SAC, and for when ballistic missiles would become operational.

In 1956 Defense Secretary Wilson assigned the responsibility of Intercontinental Ballistic Missiles (ICBM's) to the Air Force: USAF had sole responsibility for operational employments of the ICBMs, and SAC assumed the primary role in USAF's future ICBM program.

In 1958 three ATLAS D squadrons and one SNARK squadron were activated in SAC. None were equipped with missiles. But because of exhaustive and successful testing in 1957 the Air Force's strategic missiles developed to where they could make the transition from Research and Development to the operational stage. The 1st Missile Division was located at Cooke AFB (renamed Vandenberg AFB in October), California. It was transferred from the Research and Development Command to SAC. Included in the transfer was the 704th Strategic Missile Wing, which had a dual mission of training missile crews for other units and attaining operational capability with the ATLAS ICBM.

Because of the establishment of the Ballistic Missiles Committee years before, the Hq SAC staff was able to take these new responsibilities in stride.

Computer Operations

In the early days of SAC, Operations Analysis calculations were made using mechanical desk calculators, slide rules, and graph paper — high-speed electronic computers were not yet available to SAC. Even mechanical calculators will fail, as when a hamfisted operator presses two number keys or two function keys simultaneously, and there were opportunities for brilliant mathematicians to "repair" them. Unfortunately brilliance in mathematics bears little relationship to mechanical aptitude. After one of our calculators which had been jammed was seriously damaged by an analyst trying to use the only tool at hand (a beer-can opener), I levied the edict that, henceforth, cost of repair for a malfunctioning calculator would come out of the salary of any analyst caught trying to repair it. That edict didn't quite solve the problem, but it helped.

When a massive Electronic Data Processing (EDP) system was finally installed and operational in SAC headquarters, the office of Operations Analysis was permitted access for more rapid or better solutions to problems of the command. Although Operations Analysis had devised techniques for solving problems using its more primitive means, the availability of EDP opened the door to faster, more accurate, less labor-intense solutions to many really complex problems.

For example, knowledge of bomb trajectory after release was a necessary input to solution of recurring problems relating to time-of-fall, safe separation (of bombs from, and bombers in, the same formation), escape maneuvers, fuzing accuracy, and so on. In order to calculate bomb trajectory manually in a reasonable time, analysts had had to make many simplifying assumptions and approximations. With EDP they could more rapidly calculate with much greater accuracy bomb trajectory — now as a specific function of bomb ballistics, weather conditions (pressure density, precipitation, winds), target elevation, and bomb release conditions (bomber altitude, airspeed, drift-angle).

Unfortunately, when the first calculation was made with the EDP equipment the resulting trajectory data were remarkably different from results which had earlier been obtained manually. Since the analysts couldn't believe they had been so wrong so many times, they carefully recalculated manually — as best they could — the specific trajectory that had just been calculated on the EDP equipment. Again, the answer was like the old answers and markedly different from the EDP results.

Armed with their manual result, the analysts challenged the director of Data Processing. They asserted that the EDP equipment had not been adequately debugged and tested. The director had his staff rerun all the standard test programs to confirm that the system was working properly. All checks confirmed the system's accuracy.

The Operations Analysis staff studied the standard test programs and discovered that there were EDP functions that were not exercised by the standard tests. They devised a simple program to exercise those capabilities and discovered that the EDP equipment was malfunctioning in those areas. The supplier was called for maintenance and calibration. When that was completed, the analysts reran the trajectory program and, this time, obtained a solution very like their manual solution.

The sequal was that thereafter the Operations Analysis trajectory program was included in the director's standard tests — and Operations Analysts were welcomed users of the EDP equipment.

Crank Letters

General LeMay received many letters on many subjects from the public. He did not classify these as "crank" letters until the ideas in them were assessed. He always believed that no one had a monopoly on brains, and he did not want to miss a possibility. It became my chore to evaluate these suggestions and prepare a reply for General LeMay's signature. I took one letter to the chaplain and said, "Certainly this is within your purview and responsibility."

Once, by some freak of circumstances, when General LeMay picked up his outside-line telephone he found he had a caller who had somehow circumvented the normal filtering of incoming calls to CINCSAC. The caller explained that he was an inventor who had developed in his basement laboratory an explosive mixture more powerful than anything yet known to man. Despite his continued babbling, General LeMay managed to get his address. The next day Dan Johnson and Stan Lawwill drove out to the address and rapped on the door — the bell seemed not to work. After a considerable time the door was opened a crack — as far as the security chain still in place would permit — and a middle-aged man, unshaven, cautiously peered at them. After they identified themselves and showed their SAC passes he opened the door. He explained his filthy clothes and the dark house when he told them he was repairing his furnace, which had failed during the night; because he was doing electrical work, he had pulled the main fuze.

As he tried to show them around his cluttered basement with a dim flashlight, Stan suggested that he complete the furnace repair; he could tell them about his invention as he did. Actually, they were quite uncomfortable with this strange character, especially in his pitch black basement with only a flashlight that might go out. They wanted the lights on as soon as possible.

As he worked on the motor that pumped the furnace oil he told them of his long list of inventions — none of which had real commercial value. But now he was nearly through with development of a fabulous explosive. Actually, he explained, his new material wasn't really an explosive — it was an "implosive." He had put together a mixture of gunpowder and metallic ingredients which had the remarkable capability, when detonated, to cause any material "within reach" to disappear, leaving a void in space. There was none of the usual ejecta after an explosion — the material things affected by his implosive just ceased to exist! His alleged violation of the basic laws of physics did not make Dan or Stan feel any more comfortable.

Finally, the furnace was repaired and the lights were on. He showed them his laboratory, which was, in fact, a battered workbench in one corner of the basement. On it were several capped glass jars and several mortars and pestles. Although he would not tell them what his mixture was composed of, he was quite willing to detonate a few grains so they could judge its power for themselves. When they demurred, he insisted on showing them the bushel-basket-sized hole in his backyard which, he said, was a result of his detonating just a pinch of the mixture. As they backed away, promising to be in touch, he told them of the hole the size of his basement in the bank of the Missouri River near the airport where he had detonated just teaspoonfuls of his mixture. Only when they had driven around the corner did Dan and Stan relax.

They reported to me that they judged they had wasted their time spending an hour or so with that strange little character who, in their judgment, was discovering what the Chinese have known for centuries: Mixing various metals with gunpowder has exotic effects on detonation.

As Stan and I entered General LeMay's office to report on the visit, General LeMay said to me, "I hope you were careful." I replied, "I was; I sent Stan!"

Chapter 7

Strategic Air Command Nuclear Capability

Nuclear Weapon Development

Herein I am listing some famous physicists, each of whom made unique and major contributions to the development of atomic and nuclear weapons. In naming a few one runs the risk of not doing justice to others. I am guilty of that. However, those I name were indispensable, and you will see the reason for my naming them.

The following physicists, each making crucial contributions to the development of our first atomic bomb, were all born in Budapest, Hungary:

Theodore von Karman, 11 May 1881 — 6 May 1963.
Leo Szilard, 11 February 1898 — 30 May 1964.
Eugene Paul Wigner, 17 November 1902 —
John von Neuman, 28 December 1903 — 8 February 1957.
Edward Teller, 15 January 1908 —.

Teller, von Neuman, von Karman: Readers of this book know some of the contributions made by these men. I know Teller intimately. I knew von Neuman well and von Karman slightly.

Wigner: With Szilard and Einstein, he interested Alexander Sachs in the possibility of achieving a controllable nuclear reaction. Sachs persuaded President Roosevelt to appoint an Advisory Committee on Uranium. Lyman J. Briggs, Director of the National Bureau of Standards, was chairman. They met in 1939 and 1940. Wigner was at Princeton for many years.

Szilard: With Walter Zinn, he proved the possiblity of self-sustaining nuclear fission in 1939. With Fermi, he directed the Chicago experiments which resulted in the first nuclear chain reaction on 2 December 1942.

Albert Szent-Györgyi was born in Budapest on 16 September 1893. In 1937 he won the Nobel Prize. In this country he worked at the Mayo Clinic and at Woods Hole, Massachusetts.

At Sandia Corporation, Albuquerque, late 1940s. The "Little Boy" was the name for the model of the nuclear weapon that was detonated over Hiroshima, August 1945.

I have listed these Budapest-born scientists because of a thought that has fascinated me for a long time. I wonder what the world would be like today if the Hungarian government had been one that they could have supported, and they had all stayed in Budapest. They were all of the same generation, with a spread of only ten years in the births of four of them. To a lesser degree the same thought pertains to Italy and other countries.

Enrico Fermi: To represent scientists born in other foreign lands who made their contributions in the U.S., a few words about Fermi. He was born in Rome, Italy, on 29 September 1901. He died on 28 November 1954. My very first graduate in Physics from Whitman College, Harold Argo, went to graduate school and got his Ph.D. under Fermi at the University of Chicago. He was in early nuclear development and has been involved continuously since at Los Alamos Laboratory.

Nuclear Target Systems

I remember when SAC's atomic delivery capability was one wing, the 509th, with its Silverplate B-29s. My people developed the safe delivery flight paths. When the results of an atomic test did not agree with the theoretical calculations of Los Alamos, we could not say, "How interesting. We'll have to include something on that in next year's tests." SAC had to be ready to go today.

For three or four years SAC could not go to war without one of my men along. Initial fuzing of atomic weapons was extremely complicated, and knowledge of atomic weapons effects quite sparse. This need was almost forgotton once, when General Rosie O'Donnell took a bomb wing to the Far East during Korea. One Saturday morning I had to tell one of my men, Ed Winger, that he was leaving for the Far East at noon for an indefinite period. Remember, we were all civilians.

For years I had blanket travel orders. They permitted me to travel anywhere in the world, at any time, by any means. I also had orders from the Department of State to admit me to any place in the world where normally United States citizens would be required to have a passport, or passport and visa. The orders were signed for the Secretary of State by Mrs. Knight, who gained fame as the long-time Chief of the Passport Office. Her fame came mostly from passports she refused to issue.

I took copies of each set of orders on any temporary duty or vacation trips. Each also was reduced in size, sealed in plastic, and carried in my wallet. In the event of an emergency I had to get where needed as soon as possible. Passports and inoculations of all of us were kept current.

An organization (it would be prudent not to name it) told people in the Pentagon that General LeMay was sending his atomic crews to suicide. My people had determined the safe delivery flight paths. We learned about the accusation. General LeMay, General Power, Maj. Gen. J. B. Montgomery, Stan Lawwill, and I spent an entire weekend going over our Emergency War Order (EWO). We finished about midnight Sunday night. A long message was prepared to the Chief of Staff, USAF, then General Vandenberg. It said, "I am doing this in the EWO. Suggest you personally get informed." Then General LeMay said *our* weekend certainly had been shot: whom else could we bother? He answered his own question by directing that a copy of the message be sent to the Joint Chiefs of Staff.

That message caused the midnight oil to burn for three or four months in the Pentagon. When the effort was finished and about to be briefed to the highest echelons in the Air Force, a most unusual thing happened. SAC was invited to send representatives. Lawwill and I attended. Our position was completely vindicated. If SAC had had to go to war, calculations showed that if the other organization's recommendations had been used, SAC would have been less than 50 percent as effective as it was in our EWO. The safety of the flight paths was confirmed.

We knew it was extremely important to get test data to validate theoretical calculations. Except for TRINITY, which demonstrated the A-bomb, the first atomic test in the ZI was RANGER. The military request to participate had been disapproved. We went to Dr. Gaelen Felt, the test director, and on a scientist-to-scientist basis convinced him that SAC could participate on a noninterference basis.

Stan Lawwill and Paul Hower selected the Air Force camera to be used. They supervised and participated in its installation in the aircraft. They were aboard the SAC aircraft in the test. When they landed at Kirkland AFB, a small aircraft was waiting to take them to Los Alamos, where they personally processed the film. The pictures turned out to be remarkable. But it took us two years to get copies, because we did not have a need to know.

We had a very simple criterion for the credibility of our delivery procedures — would we be willing to fly in the aircraft? We not only were willing but actually did fly. When the Alsop brothers wrote that a B-36 was flopped on its back in a Pacific Proving Grounds Test, we knew what happened, as one of my men, Martin Schwartz, was aboard.

I "gave" the B-36 to the testing program without checking with Hq SAC. Not only the B-36 but a flight crew and a ground maintenance crew for six months. I hurried back to Omaha to tell General LeMay what I had done. I wanted to tell him personally, before anyone else, so that if I was wrong, or he disapproved, I would be the first to tell him and incur his wrath. He supported me 100 percent.

Can you imagine a civilian today doing something

like that? It demonstrated not only General LeMay's insight, but his total trust in me

* * *

At Hq SAC we were given the responsibility to develop an atomic test program in the Aleutians. It was not conducted, but during its development I asked one of my men if he would be willing to fly on a particular aircraft.

"No."

"Why?"

"Cowardice."

* * *

I remember when our target system was the 21 or 22 largest cities in Russia. The reason was because we had 25 atomic weapons in our stockpile, and we thought we should honor Moscow with several.

As the stockpile grew, the target system changed. For decades the target system has not been cities or population per se, but military targets and governmental control centers. Unfortunately some parts of cities would be destroyed and some civilians killed. But this is the choice of the Soviets, who would cause this to happen only if they started a nuclear war.

"City busting" has not been the primary target system for decades. That is why I regretted the political move in 1980 by President Carter to have Secretary of Defense Harold Brown announce that the target system had been changed from city destruction to military targets. The Joint Strategic Target Planning Staff (JSTPS) is a joint Army, Navy, Air Force, and Marine organization that was established on 16 August 1960. They select and recommend each Designated Ground Zero (DGZ) in detail to the Joint Chiefs of Staff. Each must be approved by the Secretary of Defense and the President.

Such a public release, with its timing, permits only one of four conclusions. Either Brown and Carter weren't listening to the presentation of the JSTPS, or neither of them understood it, or they didn't hear it but sent their proxies, or it was a deliberate political move.

Another example of political moves was the release to the media on the Stealth bomber, at about the same time.

These are only two examples of releases which resulted in 18 former senior defense officials' urging Defense Secretary Brown not to "tolerate any further revelations of sensitive hardware, policies, or intelligence sources for the President's political advantage." The signers included three former Secretaries of Defense, two former Chairmen of the Joint Chiefs of Staff, former USAF Chief of Staff General John Ryan, and former Commander-in-Chief of the Strategic Air Command, General Bruce Holloway. The 18 officials said, "Secretary Brown should refuse to politicize an office that, throughout the 35 years of its existence, has remained essentially above the political fray."

Safe Nuclear Delivery Flight Paths (General Ramey's Problem)

A major problem in Operations Analysis is that when a client says, "I have a problem; this is my problem, . . ." he is almost always right in his first statement and frequently wrong in the second statement.

One day General Roger Ramey, commander of the Eighth Air Force, then our only atomic-capable unit, came into the Hq SAC Operations Analysis Office. He said, "I am concerned that the yields of atomic weapons are being increased so much that one day my bomber will be blown out of the sky by the weapon it has just released. So I would like for you guys to devise for me that post-bomb-release aircraft maneuver that will get my bomber as far away as possible from the detonating bomb." For security reasons, we named the project "General Ramey's Problem." That seemed like a sensible request — surely the farther away you are from the point of detonation, the safer you are. Surely!

Not so! When we dug into the problem, we found that each of the weapon's effects — nuclear radiation, thermal radiation, blast overpressure, and gust — had to be considered separately. Their effects on crew and bomber are quite different and, in fact, occur at quite different times. The maneuver we finally devised answered the question, "What post-bomb-release maneuver will minimize the worst of the weapons effects on bomber and crew?" Needless to say, that maneuver provided the maximum safety to bomber and crew.

We devised the maneuver experimentally. First, after considerable research and much pressing, we obtained from the Boeing Aircraft Company assurances that there was no airborne maneuver which could damage a B-29 at indicated airspeeds less than N knots (N was then classified). N exceeded most bomb run speeds. General LeMay assigned a B-29 to us and a check crew from the 509th Bomb Wing at Fort Worth, Texas. We installed a gun-sight-aiming-point (GSAP) camera behind the copilot's seat to record the instrument panel during trial maneuvers. We used the Fort Worth RBS site to record the ground track of the maneuvering bomber.

With the B-29 at normal bombing altitude over the RBS site, at the simulated bomb release signal the pilot attempted to rack the B-29 into a steep bank to the right. Bombers are monstrous machines and respond slowly. Stan Lawwill, on the ground watching how slowly the RBS recording pen was tracking out the deviation from the straight line ground track at simulated bomb release, would shout "Harder!" into the radio mike.

As can be imagined, between flights our discussions with our crackerjack flight crew were somewhat animated. Their position was that bombers, especially the B-29, were built to fly straight and level (and with dignity!); if you want extreme maneuvers, get a fighter-interceptor! After full explanation that the safety of all SAC bombers and crews might someday in the not too distant future depend on the maneuver we were trying to develop, we all agreed to try again.

This time both the pilot and the copilot would be on the controls trying to horse the B-29 over into a steep bank to the right. It worked! Our RBS plot showed the B-29 changing direction at a speed thought impossible. The flight crew then challenged Stan to leave his plotting board on the ground and fly one of the "breakaways." He still has a piece of GSAP film that shows him in the pilot's seat and the B-29 instrument panel with the artificial horizon indicating a bank to the right of more than 90 degrees!

For years thereafter whenever Stan happened to meet one of the members of that test crew, it was old home week. Drinks all around and telling everyone within earshot of the long-hairs from Operations Analysis, Hq SAC, who had convinced the blue-suiters that bombers are maneuverable (we had done similar studies on B-50s and B-36s) — "and isn't it a good thing? Our tails depend on it!"

Breakaway Maneuvers

The question remained: For a weapon of yield K kilotons, or later M megatons, will the safety provided by breakaway, the maneuver devoloped in answer to General Ramey's problem, be adequate, or would SAC still lose aircraft and crew — just not so horribly?

In an effort to find an answer to that question, Operations Analysis laid on a research project which had two major components: (1) weapons effects, as functions of weapon yield, bomb burst time, and distance from the burst point, and (2) vulnerabilities of both aircraft and crews.

The Armed Forces Special Weapons Project (AFSWP) had sponsored and was sponsoring several theoretical research projects to provide information on the first component. Unfortunately, the theory of nuclear explosions was not yet very sophisticated, and acknowledged experts differed widely in their opinions. Worse, there were not yet any real empirical data relating to effects on aircraft in flight. Naturally, the emphasis in early tests was on learning about weapons effects on targets. Operations analysts visited all the recognized experts (there were not many), were frequent visitors to AFSWP, and attended every conference on the subject. Operations Analysis maintained a notebook of current "best" opinions on weapons effects as a basis for calculating survivability of bomb-drop aircraft.

At the same time, levies were placed on the research laboratories at Wright Field, Dayton, Ohio, to determine the resistance of aircraft components to weapons effects. Since the B-29s and B-50s were well along in their expected service lives, the labs first tested the B-36.

Operations Analysis, SAC, was delighted, despite the long delay time, to receive the first data from the labs on limits to wing-loading for the B-36. SAC analysts had recurringly "bugged" the labs to deliver to SAC whatever data they had reduced. SAC just could not wait for completion of the total test and report writing. SAC operational plans were updated continually.

Then there came a terrible shock! According to Operations Analysis calculations, using the most conservative estimates of weapons effects, with the new laboratory data, no B-36 would survive the detonation of the weapon it was then scheduled to deliver — to say nothing of programmed weapons with markedly increased yields.

SAC analysts immediately visited the aircraft laboratory at Wright Field and talked with the staff that was running the resistance tests. They discovered that the levies from aircraft structural data which had been initiated by Operations Analysis, SAC, but processed through proper channels (Hq SAC to Hq USAF to Hq Air Research and Development Command to Commander, Wright Air Development Center, to the aircraft laboratory) had been so "sanitized" somewhere en route that the test personnel had no understanding of the purpose of the test they were conducting.

We learned two additional pertinent and important facts on that trip. First, the laboratory had found the limit on wing-loading by stacking sand bags on top of a B-36 wing secured in a jig. When the first permanent deformation of the wing was detected, the weight of the sand bags was divided by the area of the wing to determine wing-load limit. Engineers now call that procedure static testing. But we needed dynamic testing results. Not many bombers deliver nuclear weapons while sitting on the ground. Dynamic wing-load limits are considerably higher than static limits, because in flight some of the stress on the wings and fuselage is relieved, as compared to sitting on the ground. All agreed that the relationship, if any, between wing response to sand bags on the ground and an atomic detonation while in flight was completely unknown. Second, we learned that the "limit" values are not the values at which the component is destroyed. That is called the "ultimate" value and is generally about 50 percent higher.

As can be imagined, we followed subsequent testing at the aircraft laboratory both personally and closely. We discovered, among other things, that some of the aircraft test designers did not really understand the basic principles of aircraft design. For example, after testing, the laboratory reported that the B-36 tail had a greater tolerance for gust loading in the positive direction (upward) than in the negative direction (downward), and gave the values of these limits.

In disbelief, aerodynamicists from my office called the chief engineer at Convair, Fort Worth (designer and manufacturer of the B-36) and confirmed that these findings made no sense. The theory of empennage design was quite mature, and Convair had deliberately designed to minimize weight (so there could not be excess strengthening in one direction). Armed with this information we persuaded the aircraft laboratory to reexamine the test concept and the test data. The laboratory concluded that test design was solid but they found an error in data reduction. The corrected values were essentially the same up and down.

After several years and many tests, Operations Analysis, SAC, probably knew more about aircraft response to atomic detonations than any other group in the world. The possible exception was the Computer Group at MIT, which later spun off as a private company. Both Rod Smith and Marty Schwartz, the aerodynamicists referred to earlier, had obtained Sc.D.s from MIT, each had been Assistant Professor of Aeroelasticity at MIT, and they knew the principals in the Computer Group and exhanged information with them on a regular basis.

Recurringly, throughout this research project CINCSAC and his staff were given status reports and our recommendations for changes to operational plans whenever indicated by the best information then available. Both the command and Operations Analysis respected the opinions of experts in the area. There is no substitute for knowing where to go to get the best information available — and having already established the rapport that would stimulate immediate response when it was needed.

Stockpile Briefing to Air Council

As seemed to happen frequently, SAC was asked on a Friday afternoon to comment on an Air Staff position recommending to the Air Council (the highest decision-making group in the Air Force) how the available (limited) nuclear material should be packaged for the atomic weapon stockpile. The possibilities ranged from large yield (relatively efficient) bombs to small yield (relatively inefficient) artillery shells. The Air Staff had reached a compromise position with the Army Staff which both would recommend to the Atomic Energy Commission, if their bosses agreed.

General LeMay asked me whether we could have an input by briefing time, 0900, on Monday. Characteristically, I said, "Sure." The philosophy which underpinned our great success and ready acceptance by the SAC Staff in those days was, "It will be a better decision with an Operations Analysis input, regardless of the time available for study." So several of us, under Stan Lawwill's direction, spent the weekend doing another analysis of how optimally to allocate scarce materials in order to maximize some reasonable Measure of Effectiveness (MOE). For that study we used as our MOE assigned targets destroyed per SAC aircraft attrited. The study was done, the briefing charts were made, and Stan was rehearsing his script when they took off from Offutt about 0400 on Monday, bound for Washington with General LeMay at the controls. They were headed for a confrontation with the Air Staff — our study did not at all support their compromise position.

At the Air Council meeting, with Chief of Staff General White and Secretary Symington in attendance, Col. Bob Greer briefed the stockpile composition recommended by the Air Staff. He stressed the difficulty and the long time required to reach a position which the Army would support.

After a brief introduction by General LeMay, which stressed the importance of using wisely our limited nuclear resources, Lawwill briefed the SAC position (the Operations Analysis study). When he finished there was very little discussion. Secretary Symington finally said these things: To General White, while pointing to our briefing charts, he said, "Tommy, that's the Air Force position; sell it to the Joint Chiefs." To General LeMay, Secretary Symington said, "Curt, I see you wisely did not put the SAC logo on your charts. We'll use them, as is, except for putting 'USAF' on each."

As they were leaving, General White observed with some chagrin to General LeMay that six months' effort by the Air Staff, studying and log-rolling with the Army, had just gone down the drain because of a weekend's activity in SAC. General LeMay responded, "Tommy it's like this: when you kiss a goose's ass, you have to be quick or you get feathers in your mouth."

Then as he put his arm around Stan Lawwill's shoulders, he added, "My staff can do better work on any weekend than yours can do in any six-month period. Besides, I don't have to get along with the Army Chief of Staff."

The War Plan vs Moscow

Because Moscow is the political center of the USSR and is of special cultural significance, the guidance to SAC war planners from the Secretary of Defense and the President, through the Joint Chiefs of Staff, was "Extreme damage to designated targets. Level Moscow if you can." In the early days when the United States stockpile was limited both in numbers of atomic weapons and in their yields, several weapons were assigned to Moscow, even though a single weapon of lesser yield had devastated each of Hiroshima and Nagasaki.

At a staff meeting when a recent maneuver was being briefed, after hearing of one foul-up after another, General LeMay said, "My God! If we do that poorly just in practice, how bad will the Moscow mission be if the whistle blows?"

I picked up the clue and laid on an Operations Analysis study of the Moscow mission as then in the War Plan. Stan Lawwill was director. Instead of assuming an average value for each of the probabilities involved in the calculation, we picked a value for each probability from the low end of what seemed to be a reasonable range of possible values. The values were, then, our pessimistic views of such probabilities as: each assigned aircraft being in commission on launch signal; aircraft not aborting after takeoff; bomber and tanker achieving rendezvous; transferring required fuel; the bomber surviving area and local defenses; navigating to the target area; identifying the assigned target; releasing the bomb accurately; achieving the right height of burst and expected bomb yields on detonation; and the accuracy of the target hardness used in war planning.

This was in the days before computer proliferation: we used graph paper and slide rules. Because there were a large number of designated targets in Moscow and several assigned weapons, each with its own DGZ, the problem of calculating the damaging effect of each weapon on each target was laborious. Ed Winger was not yet cleared for weapon information and Stan was too busy to do the whole job himself, so he "sanitized" the calculations and had Ed do the work. Stan plotted out on a blank sheet of paper two sets of points, one numbered 1, 2, 3 . . . and the other A, B, C. . . . These were, respectively, the Moscow target locations and the weapon DGZs from the War Plan, though Ed didn't know that. Stan provided Ed with a set of measuring devices labeled A_i, B_i, C_i . . . $i=1, 2, 3 \ldots n$, each with a weird scale (which actually represented the probability of target damage as a function of the distance between each DGZ and each target, though Ed did not know that either). Ed measured and recorded in matrix form the "distance" between each numbered point and each lettered point. From these numbers Stan calculated the expected damage if the bombers reached bomb release point. These numbers were then multiplied by the probabilities of the bombers reaching bomb release point and summed to give the "expected value" of the Moscow mission. As expected (since we used pessimistic values), the expected value of the mission was far from acceptable.

Stan then plotted out another pair of point sets having the same numbered points (the targets) and more lettered points in different locations (i.e., more bombs with different DGZs). The calculations were repeated, and from the increase in expected value he could better locate the lettered points for the next calculation. Finally, we had a set of weapons and DGZs which yielded an acceptable expected value for the Moscow mission.

We prepared a briefing, I reminded General LeMay of the motivation for the activity, and Stan briefed the study to him. As Stan proceeded from one flip chart to the next, showing the Moscow mission looking poorer and poorer (as successive probabilities were being included), General LeMay looked grimmer and grimmer and his cigar stub took a fierce chewing. When Stan completed the analysis and showed General LeMay the expected value, assuming pessimistic parameters, of the Moscow mission as planned, General LeMay was clearly aghast that it could be so low. Before he could comment, Stan said, "Wait! I'd like to show you the results of the same analysis for another set of weapons and DGZs."

Stan then briefed the analysis of the best set of weapons and DGZs which he and Ed had devised. When he finished, General LeMay turned, pushed a button on his squawk-box and said, "Monty, come here!" (Maj. Gen. J. B. Montgomery was SAC's Director of Operations.) He told Stan to close up the flip charts and then repeat the briefing, without change, for Monty.

As soon as Monty was seated, Stan went through the charts one by one, perhaps a bit more deliberately than before, being very much aware of Monty's increasing uneasiness under General LeMay's baleful stare. When Stan finished the Moscow mission as in the War Plan, General LeMay said, "Wait!" He let Monty squirm for a while, shrugging off his attempts at explanation, then said, "Go ahead."

When Stan completed the second briefing, Monty was much relieved, and General LeMay said, "Monty, that's our planned Moscow mission as of now!"

Stan Lawwill stated, "Thus did two civilians under a perceptive civilian boss influence the SAC War Plan."

Weapon Safety

Because in the early days atomic weapon development was more an esoteric science than engineering, it was natural that the Operations Analysis Office should acquire responsibility to stay abreast of changes and to evaluate for SAC their implications. To do this required frequent trips to Sandia Corporation, Albuquerque; Los Alamos Scientific Laboratories, also in New Mexico; and Livermore Labs in California, as well as the test site in Nevada.

Whenever there were advances which significantly affected SAC plans or operations, the Operations Analysis Office would prepare a briefing for concerned staff members at SAC Hq and, at their direction, concerned field units. The reception of, and reaction to, these briefings were generally so outstanding that the command dubbed them "OA dog and pony shows." Just as dog and pony shows used to attract everyone in the neighborhood, the Operations Analysis briefings usually played to filled auditoriums, and frequently had to be repeated for the overflow.

As would be expected, Hq SAC had made plans for a survey by headquarters personnel, including operations analysts, of any accident with an atomic weapon while SAC was responsible for it. It was ineveitable, despite the greatest care on the part of all involved, that there would be accidents with atomic weapons as SAC maneuvers with them became more frequent and more routine.

Such an accident occurred during the afternoon of 11 March 1958 when an atomic weapon was inadvertently dropped (unarmed) from a B-47 on a training mission near Florence, South Carolina. Because the survey team could not arrive at Florence during the daylight on the eleventh, departure from Offutt was delayed until 10:30 p.m. The team, including Lawwill, landed at Florence airport at 0450 on the twelfth. As had been arranged, an Air Force bus was waiting at the airport. The driver had already learned how to get to the bomb-impact site. Equipment and personnel were transferred from the SAC VC-97 and were driven to the farm on which the bomb had landed, arriving just at daylight.

The team staked off the crater area — there had been a high-explosive (non-nuclear) detonation on impact, resulting in an impressive crater — and monitored the ejecta from the crater with their radiation meters. No radiation hazard existed. After agreeing to pay for minor damage to a fence and a fruit tree, taking a statement from the farmer and his family, and assuring them that their most serious hazard was the danger of falling into the hole left by the detonation, the team boarded the bus and returned to the Florence airport. Takeoff for Offutt was at 9:00 a.m. Immediately after takeoff Commander SAC was advised of the survey team's activities and findings.

One of the team's findings was that the alpha radiation meter in their equipment was ill-designed for field survey work. Immediately upon returning to SAC Headquarters, Lawwill called experts known to him at Headquarters, Armed Forces Special Weapons Project, at Kirtland AFB and, through them, managed a transfer of two alpha meters from Sandia Corporation (next to Kirtland). The next day, Lawwill flew to and from Kirtland, bringing back both meters after careful briefings and demonstrations by Sandia and AFSWP personnel. As he turned the meters in to the custodian of the survey team's equipment, Lawwill said, "I'm told we now have the best two alpha meters in the whole country for survey work; it's nice to have good friends in important places."

As an aside, among the goals which I regularly enumerated was that I wanted each Operations Analysis staff member to be on a first-name basis with the foremost experts in his assigned area of expertise. Most of them were.

Command Post Exercise (CPX)

In order to assure that, if war came, the NATO forces would effectively support one another and contribute optimally to the SAC SIOP (Single Integrated Operations Plan), Command Post exercises were held annually. These simulated wars exercised the communications links among the Allies and tested the deployed command staffs. In the early days, only SAC had an atomic delivery capability and the SAC command staff always deployed with it a member of Operations Analysis who was skilled in selection of fuze settings for atomic weapons. These varied, of course, with intended target, weapon yield, and release conditions. These parameters could not be known in advance, since the war game scenario varied from year to year, so it was important that the fuze setting could be calculated on call and without delay. In those days, fuzes were set when the weapon was assembled and could not be changed in flight.

At one CPX in Paris the following events occurred. General LeMay and his command staff, including Stan Lawwill of Operations Analysis, arrived in Paris just in time to convene with the NATO staffs on Tuesday morning. In addition to the thrill and satisfaction of participating with General LeMay in the CPX, where they shared in the respect and admiration accorded the general, the SAC staff anticipated the end of the exercise and "doing the town."

In that particular CPX all of Tuesday, Wednesday, and Thursday were spent in the exercise of the conventional (nonatomic) forces, including Air. Finally on Thursday night the Allied High Command agreed that the simulated battlefield situation was so bad that SAC's atomic forces had to be used. The SAC atomic battle staff had not been just standing by for three days. As the war game progressed they had been planning for, and continually updating, the use of SAC's atomic forces. Furthermore, General LeMay was periodically updated on the tentative atomic plan and had provided guidance on emphasis and priorities.

When the request for nukes came, the SAC staff was ready; General LeMay was briefed; after minor changes, he accepted the plan and loading of weapons, launching of bomber and tanker forces, rendezvous of bombers with assigned tankers, refueling, massing at entry points, penetration of defenses, bomb release, and return to recovery points — all in just slightly more than (simulated) elapsed flying time.

By late morning Friday the simulated results were

tabulated and there was unanimous agreement that the war was over; the SAC atomic forces had been clearly decisive. You can imagine the exhilaration of the SAC battle staff as it assembled for busing back to the hotel rooms, which had been unused except for quick showers and shaves and occasional catnaps for three days. After a nap Friday afternoon, Paris was at their feet!

The elation of the staff was shortlived. As he climbed on the bus, just after he commended the staff for an excellent exercise, General LeMay said, "We might as well go over to England this afternoon; there's nothing to do in Paris on a weekend. Wheels up for Ruislip at thirteen hundred hours." Ah well; London isn't the worst town in the world — and besides the women speak English there.

Fuzing Problems

In the early days atomic bombs were far different from the kinds of munitions that the military were accustomed to. After all, they were designed by research scientists rather than munition engineers. Research scientists, especially nuclear physicists who were pressing the edge of the state of the art, had very little feeling for the problems of the military operators. Complexities of design were no consideration. After all, the fact that an atomic weapon would detonate at all was nearly a miracle; why bother with niceties such as ease of assembly and use?

As might be expected, the early weapons were more laboratory devices than munitions. The bomb arming and fuzing train, for example, was a series of devices all of which had to function in sequence in order to detonate the bomb.

One Saturday evening when my wife and I were entertaining the operations analysts at our home, there came a call from the SAC duty officer saying Stan Lawwill and I should report to General LeMay at once. A delegation from the Armed Forces Special Weapons Project at Kirtland AFB had arrived to brief CINCSAC on the very serious fuzing deficiency AFSWP had just deduced from their latest test series. The conclusion of their briefing was that the weapons then available ccould not be used in the SAC war plan. Clearly, such a conclusion was not acceptable to CINCSAC!

General LeMay dismissed the visitors with the cynical observation that he and Lawwill could build a better fuze using a two-dollar watch. The SAC staff then discussed the effects of weapon degradation on the war plan. General LeMay assigned to Stan and me the problem of deciding how best to change fuze settings in order to minimize the degradation of weapons effects while adequately compensating for fuze deficiencies.

* * *

A timer in the early atomic weapons counted the time from release to actuation of active fuze components. As had been recommended by Operations Analysis, SAC was using that timer as a sort of ceiling over the height of burst, based on the presumption that the Russians would try to predetonate the weapon (so high above the target to reduce target damage). Because early atomic bombs did not have impact fuzes, it was important that the timer ceiling over the height of burst be high enough to permit full fuze actuation before the weapon impacted. Otherwise, there would only be low-order (partial-yield) detonation.

Having responsibility for determining fuze settings, Operations Analysis carefully calculated the setting for this particular timer, which could measure time to one-thousandth of a second. Such a setting could have been, for example 28.293, which meant that the bomb could not be detonated until it had been falling at least 28.293 seconds after it was released from the bomber.

Having made such calculations, Lawwill was distressed to observe a weapon assembly maneuver. In those days fuzes were set at the time the weapons were assembled. He saw that the timer had no dial or hands or knobs. It was set by a screwdriver adjustment, and the time interval had to be measured using an A-scope (a standard oscilloscope). The standard operating procedure called for running the timer to measure the interval, making a screwdriver adjustment, and repeating the above process until the measured time interval was within two-thousandths of a second of the prescribed setting on three of five successive runs. As can be imagined, that was a time-consuming process.

After returning to headquarters, Lawwill calculated the loss in weapon effects in the war plan if each timer setting was rounded to the nearest second. He persuaded General LeMay that the possible loss in target damage was very small compared to the increase in operating flexibility permitted by this significant saving of weapon-assembly time. General LeMay signed out the message which Lawwill had prepared directing all weapon-assembly units henceforth to set each of those timers to the nearest second specified in SAC's fuzing instructions.

For the next several weeks Lawwill showed clear signs of being impressed wth the great contribution he had just made to SAC's War Plan. Then he went on another weapon-assembly maneuver — obviously to relish and personally take full credit for the great amount of time he had saved the assembly team. Horror of horrors! He observed that the timer setting procedure was unchanged. The fuze was now being set to twenty-eight point zero, zero, zero! Needless to say, the next instructions which he persuaded General LeMay to sign out were much more explicit — and did achieve the desired results.

Radar Bomb Scoring

Of the fine training devices for bomber crews that

were developed during World War II, one of the best was Radar Bomb Scoring (RBS). An Army-developed, gun-laying radar (SCR-584) would track a bomber making a simulated bomb-run over a target complex in a U.S. city. Using azimuth, elevation, and range data from the radar, an automatic plotting board would record the ground track of the bomber. At simulated "bombs-away" a mark on the bomber's ground track would permit the RBS crew to quickly plot where the (simulated) bomb would have detonated.

The distance and direction to the target were then easily measured and communicated to the bomber crew. The data from the radar on bomber altitude, ground speed, and heading could be quickly converted (using wind data collected from an earlier tracking of a corner-reflector carried aloft by a Rawin balloon) into bomber heading and true air speed. These data, plus bomb ballistics (time of fall and trail angle) from bomb tables, could be compared immediately with the airborne bombardier's calculations and settings. Quite generally, after such a comparison, a second bomb-run yielded much better results — sometimes a "shack" (a direct hit). The great merit of this training technique was that the bomber and crew were operating under realistic combat conditions: the altitude, airspeed, target, and bomb ballistics were just as they would be in combat.

Operations analysts were largely responsible for development of the scoring techniques used by the RBS crews. An operations analyst always accompanied the inspection team that evaluated each new RBS site as soon as it became operational.

The first of the RBS units to become operational was "Kansas City Kitty." It was located on the levy at Kansas City (in order to minimize nearby obstructions) and the location was surveyed in by the U.S. Corps of Engineers. The RBS crew plotted the locations of selected targets (specified corners of a bridge abutment, a factory, a railroad marshaling yard, and so on), as well as the RBS site, on an AAF target chart of Kansas City. From this plot a target template was made for transfer, before each set of bomb runs, to the paper on the plotting board.

One of the most impressive results of RBS training using Kansas City Kitty was the remarkable reduction in the bombing error of the first B-29 wing to be equipped with the Eagle radar — the APQ-7. The antenna of the APQ-7 was a fixed, wing-like vane attached beneath the fuselage of the B-29, parallel to the wing; in fact, it looked as though the B-29 was modified to be a biplane.

Instead of the radar beam being swept mechanically by a rotating disk antenna, it was swept electronically (by phase shifting the signal along the fixed antenna radiators). Theoretically, the APQ-7 showed great promise for improving radar bombing accuracy. In fact, when first used by operational crews, the APQ-7 displayed disappointing results. Since the B-29 wing was scheduled to deploy soon, a crash effort was made to upgrade bombing accuracy by devoting Kansas City Kitty to that wing exclusively for three weeks. The average bombing error for the whole wing at the end of the excercise was better by a factor of nearly three than the average at the beginning of the three-week period!

An interesting sidelight of this exercise — and an illustration of the discoveries that can be made by an inquisitive mind that has the time for a broad view, for reflection, for questioning — was the analysis made by operations analyst Mark Eudey at Hq Second Air Force, Colorado Springs. He was not satisfied merely to receive data and compute the great reduction in bombing error during the 3-week exercise. In addition, he analyzed errors separately for each of the targets. He discovered that for each target there was a bias error; all biases were of about the same magnitude and the same direction. There is no physical explanation of how a bias error — for many radar bomb runs against several different targets by bombers with different headings, altitudes, airspeeds, and winds at altitude — could relate to aircraft equipment. The only other equipment involved was the RBS unit.

Since Stan Lawwill was at the unit he was asked to look into the matter. He discovered, as Mark Eudey had suspected, that there was a significant error (it turned out to be 194 yards) in the location of the RBS site, as specified by the Corps of Engineers. In fact, he discovered that the (old) U.S. Engineers' map differed significantly, especially along the river, from the AAF target chart of the same area. Of course, these differences were small compared with radar-bombing errors.

Needless to say, thereafter even surveys by the U.S. Army Corps of Engineers were cross-checked before any RBS site was declared operational.

As would be expected, when World War II was over RBS became an important training device for SAC. By that time equipments had been well shaken down; scheduling, scoring, and reporting had become routine; and there was no need for special attention by operations analysts, except when RBS was to be used for special or novel exercises (to score toss bombing, for example). As was the custom in the Operations Analysis Office whenever a newly pioneered technique or equipment became fully operational, responsibility for RBS was turned over from Lawwill (who, of course, was available for consultation as needed) to a bright young captain in Operations, Capt. Harold Stack (who retired as a major general).

After the war a new RBS equipment (the MSQ-1) was developed. Lawwill was assigned to the review team that monitored and guided the development. Among his responsibilities was assuring that the new equipment would, in fact, measure the accuracy of SAC bombing over the whole spectrum of release conditions and provide data of a diagnostic nature, so it would be possible to improve SAC's bombing accuracy (already the best in the world!).

Among the requirements established by the Air Force for the MSQ-1 was automatic (non-manual) scoring. Ford Instrument Company had the contract to develop the computer that would do the automatic scoring. The Ford Instrument Company design engineer for this project was Morton Sobell. Because bomb-release conditions and SAC's bombing accuracy were classified, the computer development was being managed at the Secret security level. This meant, of course, that the development review team had to be very careful in what it said in the presence of contractor personnel, some of whom were not cleared.

Shortly after a review team visit to Ford Instrument Company there came a telephone call from Sobell to Lawwill at SAC headquarters. Sobell asked for SAC bombing accuracy data and ballistics data on bombs. Lawwill gave him no information over the phone but said he would collect the data while Sobell was processing a formal letter of request through his security office and the Air Force Cognizant Officer. The letter never came. Morton Sobell was found guilty of espionage and sentenced to 30 years in prison.

Flash Blindness

When an atomic weapon detonates, there is a tremendous flash of light, and the thermal radiation is so intense that damage can be done to the retina of the human eye at distances of many miles. Even when the viewer is not looking directly at the explosion (so there would be no permanent retinal damage), he may experience a temporary sensation of flash blindness, lasting from a few seconds to a day.

Clearly such incapacitation would be very serious to a bomber crew flying over a target area — yet that is precisely where there will be atomic detonations. So, among others, the SAC operations analysts who participated in atomic weapons tests tried various techniques to prevent, or to minimize effects of, flash blindness. Of course, every crew member was warned not to be looking in the direction of ground zero at time of detonation. In addition, some crew members were instructed to close one eye just before time-zero. Then, a few seconds after detonation, each would open the closed eye to see whether he could see with it, or whether it was experiencing sympathetic blindness. In either case, each crew member noted the time elapsed before he could read instruments again.

Some crew members found too much light penetrated the eyelid of the closed eye so they were experiencing flash blindness in both eyes. Some of them tried to cover the closed eye with a hand. Others, finding that they needed the use of both hands, tried putting on their Air Force sunglasses (with the big, wrap-around lenses) just before detonation, and just tolerating the reduced vision in the still-open eye. Some found that sunglasses reduced the light reaching the closed eye long enough to protect the closed eye from all but a few seconds of incapacity.

Two amusing incidents relative to flash blindness:

During one of the atomic detonations at the Nevada Proving Ground, Lawwill was to be airborne at time of detonation. Just before his B-29 was scheduled to take off from Kirtland AFB, Albuquerque, in the black of night (detonation was scheduled just pre-dawn), the test director appeared to ask whether the young Navy lieutenant with him could be taken along. The answer was "Yes," and because the lieutenant was not dressed for that flight, Lawwill lent his flight jacket to him and told one of the crew members to brief him on flash blindness.

After the flight and the return to Kirtland, when Lawwill went to retrieve his jacket, he found the young lieutenant in great distress. Having found a pair of Air Force sunglasses in the jacket pocket, the lieutenant had tried to use them to prevent flash blindness. He found that on landing, an hour or so after detonation, he still could not focus his eyes even though he had been trying to read aircraft instruments beginning a few seconds after detonation. Lawwill suggested the lieutenant's sight would probably straighten out as soon as he took off Lawwill's prescription-ground sunglasses.

One of the solutions to problems of flash blindness was to cover all windows in the crew compartment with "blackout curtains" at the beginning of the bomb-run. From that time, the bomber is on "second-station," which means the radar bombadier is guiding the aircraft so the crew doesn't need to look outside. An operations analyst recurringly visited each SAC wing to brief on effects of atomic weapons, flash blindness, and so on.

When an analyst visited a wing which had just received its first, partial, allotment of blackout curtains, he found the curtains had not yet been distributed to the bombers. Checking, he found the director of Maintenance and Supply very much annoyed at the stupidity of those higher-echelon clowns who had distributed blackout curtains before they were blackened. He took from the original package and showed the analyst bright white canvas curtains that his maintenance staff had not yet sprayed. The analyst, perhaps not as gently as he should have, explained to the colonel that "black-out" was a relative word: the white curtains would prevent flash blindness by keeping out the flash of detonation — and would not be ignited by the thermal radiation, which surely the blackened curtains would be. They both agreed that the danger of fire in the crew compartment exceeded the danger of flash blindness. The colonel said, "I'll just have to requisition more sets of curtains and sweat out their delivery."

Strategic Air Command Support of SACEUR (Supreme Allied Commander Europe)

Stanley Lawwill has contributed the following report.

There came the day when there were enough atomic weapons that a few were assigned to Supreme Allied Commander, Europe. At that time SACEUR was a Britisher, not cleared for U.S. atomic weapon information. So he designated his American deputy, General Norstad, to represent him in negotiating with CINCSAC on delivery of weapons assigned to SACEUR (who had no atomic delivery capability).

General LeMay, General Cam Sweeney (director of Plans), Col. Eddie Broadhurst (deputy of General Sweeney), a colonel from the Directorate of Materiel, and I flew to Paris to arrange for SAC to deliver the SACEUR weapons. We had a fairly brief discussion with General Norstad and were directed to the Commander-in-Chief, United States Air Forces, Europe (CINCUSAFE) at Wiesbaden, Germany, to complete detailed arrangements. CINCUSAFE, after welcoming us warmly, told us that his director of Operations was most knowledgeable in this area and would make arrangements with us.

General LeMay flew on down to North Africa to visit the SAC bases there, leaving General Sweeney to settle with CINCUSAFE. General Sweeney, in turn, made Colonel Broadhurst responsible and, borrowing CINCUSAFE's administrative aircraft (a Gooneybird, C-47, which had been nicely paneled inside), returned to SHAPE (Supreme Headquarters, Allied Powers, Europe) in Paris on other SAC business.

Eddie Broadhurst, the colonel from Materiel, and I met with the CINCUSAFE director of Operations and his staff. Several things became immediatedly clear. The Materiel colonel really was not involved, so he left. The staff of the director of Operations was neither knowledgeable nor cleared for atomic information, so Eddie and I had to be careful to "sanitize" the things we said. Finally, the tentative plans CINCUSAFE had devised for the use of his assigned atomic weapons were neither very good, nor compatible with CINCSAC's war plan.

The director of Operations finally excused his staff and the three of us hammered out an agreement. When Colonel Broadhurst briefed General Sweeney the next morning on his return from SHAPE, Sweeney looked at the two of us as though we were the cat that had proudly brought the mouse into the living room. He said, "You two must have forgotton that you represent CINCSAC! I would be ashamed to take this agreement to General LeMay." He turned on his heel and left, saying he would be back next morning.

Eddie and I wondered at some length what better we should have done, decided, and called the CINCUSAFE director of Operations (who had already briefed CINCUSAFE) to tell him that we did not yet have a satisfactory arrangement. We spent much of the evening persuading him that he should relinquish more of his prerogatives — that SACEUR would be better supported if the SAC staff made more decisions for CINCUSAFE on the use of atomic weapons allocated to SACEUR.

Our prideful report to General Sweeney the next morning was crushed by his response. "You guys have missed the boat," he said. "General LeMay will pick us up at Orly tomorrow at 1700 so you have time for just one more session before I have to brief General Norstadt." So saying, he left and Eddie and I regrouped.

That day we persuaded the director of Operations, USAFE, that the science of weapons effects, target hardness, selection of optimum height of burst (fuzing), and selection of DGZ was still an art and evolving very rapidly; with each new atomic test, giant strides were being made. The SAC staff, namely Operations Analysis, was participating in the tests and helping to reduce the resulting data — therefore SAC was as immediately current as it was possible to be. Clearly, on a continuing basis, SAC could keep SACEUR's plans for use of atomic weapons in much better condition than could CINCUSAFE. The director agreed, and we so informed General Sweeney when he came to pick us up for the flight to Orly.

Enroute, at a time when only General Sweeney and I were together in the forward cabin, he said, "What's the matter, Stan? You seem distressed." I told him bluntly that I thought he had grossly mismanaged the whole deal. If he didn't trust us to make a good arrangement, he should have given us prior guidance, rather than just shrugging off our negotiations. At first he looked at me severely as though he were about to chew me out for my brashness.

Then he smiled wryly and said, "I'll tell you Doc, a long time ago I learned that I was not the smartest staff officer in the world. Did you have any idea that you could have negotiated such a deal for General LeMay? I didn't! If I had given you guidance, we'd now have a much less advantageous plan than we have. And I know General LeMay will appreciate our good work!"

He did.

Chapter 8

Strategic Air Command Operations Analysis Field Offices

While I was Chief, Operations Analysis, United States Air Force, there were 18 different Operations Analysis offices around the world. In the ZI there were offices at the headquarters and some units of the Strategic Air Command, Tactical Air Command, and Air Defense Command. There were offices in Hawaii, Vietnam, Japan, Alaska, Europe. Believing that one cannot do one's job properly sitting at one's desk, I made numerous field trips. This responsibility contributed to my 40 trips to Europe, seven to the Far East, and many to other parts of the world.

The office at Hq USAF did not control the field offices. The chief of each field Operations Analysis Office reported to the commander of the unit to which the office was assigned. The Hq office, upon request, would provide advice and consultation, and often personnel augmentation to the field units for special projects. It provided coordination, helped avoid duplication, and helped the field units take advantage of work already accomplished. It provided information and identified sources of information.

Often the larger computer capability in the Pentagon was used to assist special field unit projects. Two rather large projects were undertaken, one for the United States Air Forces, Europe, and one for the Pacific Air Forces in Hawaii. Those headquarters sent men to the Pentagon, who were augmented by personnel from my office; we arranged the computer time, helped with the model-building, and assisted in the analysis.

The Hq USAF Office arranged yearly technical coordination conferences for representatives from all the Operations Analysis offices. These meetings were held at different locations each year, in the ZI and in Alaska, Europe, Puerto Rico, and Hawaii. Papers on methodology, proper analysis tools, input sufficiency and relevance, and interpretation of results were presented and discussed.

The office at Hq SAC operated in much the same way in its relationship to the Operations Analysis offices at SAC's numbered Air Forces and at the First Missile Division at Vandenberg AFB.

Examples of contributions made by SAC field offices follow:

Hq Second Air Force, Barksdale AFB
(Reported by Ed Berdahl)

Effective 1 November 1949 the Second Air Force was established and assigned to SAC. Hq Second Air Force was activated at Barksdale AFB, Shreveport, Louisana.

In early 1950 the Second was almost exclusively involved with reconnaissance activities. Expansion and the integration of B-36s and B-50s in the command required a more balanced organization. Geography was also involved. The Second controlled units at Fairfield-Suisan AFB, California. Each numbered Air Force was assigned both bomber and reconnaissance aircraft. The Second was assigned units on 1 April 1950 in the Eastern part of the United States.

A realignment of the numbered Air Forces on 13 June 1955 gave the Second the responsibility in Southeast United States plus Texas.

On 1 July 1957 SAC assumed jurisdiction of the Sixteenth Air Force at Torrejon Air Base near Madrid, Spain. The Sixteenth commanded the 5th Air Division in French Morocco. Since the Sixteenth was not capable of taking over control of SAC operations in that area at that time, Second Air Force retained this responsibility throughout 1957.

Effective 1 January 1959 SAC realigned several bases and units between the Second and Eighth Air Forces. This realignment placed the Second in command of forces in the Central section of the United States.

In a realignment on 1 July 1963 the Eighth acquired from the Second a Titan II Wing at McConnell AFB, Kansas, and Whiteman AFB, Missouri.

In another realignment in 1970 the Second became an all-manned-aircraft command, consisting of B-52, FB-111, and KC-135 units.

After Generals Old and Sweeney became commanders of the Fifteenth and Eighth Air Forces, respectively, and after Glen Hadsell and Tex Alexander had helped Second AF analyze LEAP FROG I, a deployment to North Africa (later I helped analyze LEAP FROG II), there was a genuine desire to have a few experienced operations analysts at Hq Second Air Force, rather than depending on service from the Field Operations Division of Operations Analysis at Hq SAC.

Ed Winger was quick to volunteer to move to sunny California to serve Fifteenth Air Force, and Tom Chappelle was soon persuaded to move to Eighth Air Force Hq in New England, but there was less enthusiasm about moving to Shreveport, Louisiana. Harry Chittim was offered the chief's position, and he said he might be willing to accept if I would go with him. When it was proposed that I move to Second Air Force as Chief, Operations Analysis, the challenge and promotion potential caused me to accept the offer, and I moved to Shreveport with my family in July 1955. Dr. Fielding McGehee had been hired as a staff member in anticipation of forming the Operations Analysis office there. I then recruited E. J. Culling as my third staff member.

General Frank Armstrong, as Commander Second Air Force, believed in delegating responsibility. His philosophy was to "place the monkey on your back," and "heaven help you if you shook it off." It was an opportunity to seek and find a constructive role for us on the staff. The attitude of some of the young staff officers was, "We're already tops in the SAC family, so why do we need civilian scientists around here to advise us?"

However, key individuals such as Col. George Thorpe, Chief of Operations Plans, Colonel Stillson in the Directorate of Intelligence, and the staff in Operations Training were enthusiastic supporters of Operations Analysis. We soon were deeply involved in such projects as evaluation of electronic countermeasures missions, assessing B-47 breakaway maneuvers by using RBS radar tracks and bomb ballistics to estimate the g-load kick an aircraft would receive from its own weapon after detonation, and statistics about the risks of handling and carrying nuclear weapons.

As Hq SAC personnel like Brig. Gen. R. M. Montgomery rotated into command positions we were given tasks and consulted more frequently by the Command Section, but our bread and butter was working with the staff officer majors and captains in Operations Plans, Operations Training, and Intelligence. If we helped the operations officer draft the operations order we could then call for data recording to provide material for objective analysis which could supplement the subjective evaluation normally obtained from participants and staff observers.

One excellent example of this approach and its influence on operations and war planning was our evaluation of the KC-97 airborne refueling of B-47 bombers. On a typical mass training mission, 20 B-47s would depart on a simulated wartime profile mission in which a rendezvous with an appropriate number of KC-97 tankers would be planned prior to arrival at the programmed target distance. After the subsequent RBS run or practice bomb drop the B-47s would fly back to their home base. The distance and altitude flown would be planned to be similar to a B-47 departing from its pre-strike base, refueling, going to target, and returning to its post-strike base, which might not be its pre-strike base.

The typical air refueling would be a 40,000-pound transfer. If the B-47 got its 40,000 pounds, the subjective evaluation would conclude that this portion of the mission was 100 percent successful. When we started reviewing these missions we found that many factors could reduce the true value or "true net gain" from the 40,000-pounds refueling transfer. It was difficult to marry a slow-flying propeller-driven KC-97, going as fast as it could and using a substantial descent rate, to a B-47 which, when it was nearly full, was flying almost as slowly as it could with full thrust on all six jet engines. It was a credit to the aircrews that refueling operations were as successful as they were, and that very few serious accidents ever occurred though in the late 1950s approximately 700-800 B-47s were flying in the Second Air Force alone.

Other factors such as rendezvous problems, rough air with its associated disconnects, training level of either aircrew involved, and the need to conduct refuelings at altitudes considerably less than the desired 18,000 to 20,000 feet MSL because of weather, all tended to decrease the refueling effectiveness.

Because of these variable penalty factors we began to calculate the true "net gain," that is, the equivalent instantaneous fuel transferred to the B-47 if it had remained on course toward target at cruise altitude. This was computed to be about 33,000 pounds from a 40,000-pound transfer if conditions were near perfect, and the demonstrated normal amount of time was spent in rendezvous, hookup, and transfer; and if the penalty of descending from and climbing back to cruise altitude was considered. However, after we implemented a rather thorough air refueling data collection form and analyzed these data, we found that this distribution of net gains accomplished showed that a relatively high portion of these B-47s, 25 percent, had net gains less than 25,000 pounds from a 40,000-pound transfer.

On training missions within the continental U.S. such a shortfall could be compensated for by cutting corners on a typical triangular route, or by calling for additional air refueling while enroute back to home base. However, if such a reduced net gain was suffered on a straight "out and back" mission, many of these B-47s could not make their post-strike base as planned. Therefore we emphasized data collection on refueling where the B-47s were deployed to or returning from SAC forward bases in North Africa, Spain, or England. During this period each wing was sending three B-47s each week for a typical four-week stay overseas. Over 700 air refuelings were analyzed in depth, using computer sorting and data storage techniques.

These refueling records obtained on overseas deployment were most valuable. After reporting these data to the Second Air Force staff and to a Wing Commanders' Conference, we were asked to brief each wing so the aircrews could see some specific results, and so they could feel that they received something in return for their inconvenience in filling out the air refueling data forms. The briefings were very well received at wing level. Naturally the lead and select crews achieved better results on the average than the less experienced crews. These results did affect the war plans to the extent that post-strike bases were replanned to enable a higher percentage of B-47s to make their post-strike base with a lesser net gain based on the statistical distribution obtained during this study. Naturally, we felt we had contributed something significant to enhance the effectiveness of Second Air Force — a sense of accomplishment so necessary to a dedicated operations analyst.

When Maj. Gen. J. P. McConnell came to Second Air Force from Hq SAC to be our new commander, he asked us to study the merits of rotating a complete wing of B-47s to North Africa or Spain for a period of 90 days versus the then-current concept of rotating a few aircraft every week from several wings to a given base, to sustain the same number of aircraft on alert in the forward area. SAC had operated on the 90-day rotation in the past, but it was quite disruptive to training and support operations back at the home base, and it also meant a relatively long separation from families, which inevitably contributed to an increasing divorce rate among wing personnel.

We looked at the overall goal of keeping *x* number of aircraft on alert in the forward area, but doing so by risking the least aircraft and personnel resources in the more vulnerable forward area. Under the frequent, small-rotation concept all but a small fraction of the time spent by B-47s and their aircrews in the forward area was spent on alert. Furthermore, in the event that wartime would require the rest of the wing to rotate to the forward base, their own wing personnel would always be on hand to receive them. They would know the exact status of the base, current phone numbers of key base personnel, and so on.

On the all-or-nothing 90-day wing rotation, the wing personnel would rotate to their forward base about once every one and one-half or two years; many personnel changes would have taken place both in the wing and at the forward base during this period, so operations probably would not start so smoothly as they would with the smaller regular rotation by all wings. Our use of the primary MOE (Measure of Effectiveness) of "alert days generated in the forward area per personnel and material resources at risk" clearly favored the smaller regular rotation. After this study, General McConnell enthusiastically supported this "steady flow" concept.

As time went on it was decided that we needed to occasionally test the speed and reliability with which we could launch our alert force. Naturally one of the largest concerns was the risk of a nuclear weapon incident, so we were asked by Vice Commander, Brig. Gen. R. M. Montgomery to evaluate these risks. We dug into all the records we could find about the probability of a B-47 crash on takeoff (one in 10,000), and the risk of dropping a weapon in handling or while transporting it on the ground, and in the air. Our best estimate was that the risk of an accidental detonation of nuclear yield was infinitesimal, but we estimated, using all available history, that if 120 sorties were launched from alert posture there was a 50/50 chance that some sort of incident would occur, that is, a compromise of only one of the several safeguards, inadvertent release on the ground or in the air, an aircraft fire, or an aircraft crash.

I was sitting in the chair having a haircut when the phone rang. The barber said, "It's for you from General McConnell." When I got on the phone he said, "Hold the phone close to your ear." "OK," I said, and then he asked, "When you have a high-explosive explosion do you produce white smoke or black smoke?" I said I'd go to our office immediately and check our references and consult with my associates.

Culling came up with the sensible conclusion that if the detonation is in contact with the ground then soil is disturbed and entrained and it appears as a cloud of black smoke. If the detonation is off the surface it would appear to be a white cloud. We then learned that an atomic weapon had been inadvertently released near Florence, South Carolina, after the alert aircraft had been scrambled on an alert test. It resulted in a low-order high-explosive (HE) detonation upon impact, with no injuries to personnel on the ground, but it did verify that incidents could occur, and that the many safeguards designed into the weapons were preventing serious consequences.

Another interesting short study we made concerned the premise that "the probability of a successful launching of a B-47 from an alert posture without having a ground abort decreased with longer periods the aircraft were on alert." The standard alert procedure was to assign a new aircrew and flight chief to a specific B-47 every morning after the previous crew had served 24 hours on alert. To avoid loading and unloading the weapons and "ammo" on the B-47, the aircraft was assigned to stay on alert for various lengths of time, like three days, a week, ten days, two weeks, and even three, four, and six weeks in some unusual circumstance, as in the forward area during the Suez crisis. The common opinion among pilots was that the longer the aircraft sat on the ground the lower the probability it would pass its preflight. Nobody likes a hangar queen. As one of the few SAC operation analysts who was also a pilot I fully appreciated their view.

When we decided to look at these data we found that a nice smooth curve of the average "ground abort rate" *vs* "time on alert" was just the reverse of what we expected. The reliability of launch increased from roughly 90 percent after three days on alert to approximately 96-97 percent after four weeks on alert. After considerable study of this trend we concluded aircraft on alert really experienced quite a different environment than a day-to-day environment for other B-47 aircraft. The key difference was that the aircraft were being thoroughly inspected every morning by the new aircrew, who knew they might have to take off on short notice with only the barest minimum preflight checks before takeoff. Therefore, more of the minor deficiencies (the kind that could be worked on when the aircraft was on alert) were being found and fixed, and the quality of the aircraft condition was actually improving as different aircrews carefully preflighted the aircraft repeatedly.

Our basic purpose in Operations Analysis was to assist the staff in professional, objective evaluations of any new ideas or existing operations considered of importance to Second Air Force. One item of concern to war planners was the ability of bombers to arrive at a designated control line on time so that simultaneous mass penetrations could stress enemy defenses to a maximum extent and thus maximize penetration success. Lt. Col. Larry Connolly was Operations Officer and later Commander of the Weather Squadron based at Barksdale and serving Second Air Force. He had a concept for B-47 navigation which used the barometric pressure differences between the east and the west boundaries of the Atlantic Ocean to compute a single heading correction which a B-47 could hold during the complete over-ocean leg. The actual ground track could drift left or right at various points enroute depending on the weather pattern, but the "coast-in" penetration was much more predictable than if the aircrew took several celestial observations enroute and tried to keep on a true great circle surface route.

The reason was that sometimes aircraft would drift left on the leg before the celestial shot was made. If a drift correction was made to the right, and then wind directions pushed the aircraft to the right instead of the left on the next leg, the aircraft deviated even farther from the planned track. This sort of oscillation along track was not unusual if celestial naviagation alone was used; hence arrival at a control point was less predictable than desired, and errors always made the aircraft late. Colonel Connolly's method also simplified flight plans so much that the navigator-radar observer could make the detailed flight plan during climbout after being scrambled from alert.

When Maj. Gen. George Mundy was Commander, Second Air Force, he flew a B-47 from Goose Bay, Labrador, to a "coast-in" point in North Africa using this single-heading flight method and hit the coast within 10 miles of desired point and within five minutes of programmed time. Later a controlled test was run from Hunter AFB, Savannah, Georgia, in which one of the two wings based there flew to England by the conventional celestial navigation method, and the other wing flew using the Connolly single-heading method. The latter produced arrival times and positions that were more accurate and far more predictable; the arrivals by the old celestial method had a median result definitely on the late side, with some as much as 45 minutes late.

After Col. Steve McElroy became Commander, 376th Medium Bomb Wing at Barksdale, he requested us to study their ECM equipment and tactics. He believed they were too subjective, and generally had experienced observers grade ground-based radar scope presentations on a scale of 1 to 5 on their estimate of increasing difficulty in following bomber penetrations of defense air space.

June Culling developed a method by which he compared each bomber navigator's track maps versus time (showing check points versus actual time of arrival) to the Air Defense Command's radar teller's logs showing where they had observed penetrating bomber aircraft. He calculated the size of a mission-derived box of "error to side of actual track" and "error along track," which would include 90 percent of these error distances determined from the closest bomber position to each teller's log plot. He then classified every plot into "accurate" or "inaccurate" categories depending on whether the teller's log plotting error was less than or greater than the 90 percent box dimension.

After this sorting he could determine the accurate plots per thousand miles of bomber track. This measure of effectiveness was very useful because we could determine how well a given type of ground radar site could track bombers at the edge of a formation compared to the middle of the formation, lead bombers compared to trailing bombers, mixed altitudes compared to common altitudes — all these variables with or without chaff dispensing, or with and without active jamming, or combinations thereof.

We found this MOE would drop off with increasing distance from a radar site, just as expected. We also could see the difference in performance of different ground-radar types. Some analysts were skeptical of the quality of bomber-navigators' logs, so each navigation plot was checked by analyzing the average ground speed between adjacent check points. If a substantial anomaly was found, such as a high-speed leg followed by an unusually low-speed leg, the check point would be adjusted along the track to indicate a consistent speed between check points. On this particular test mission, over 40 B-47s met off Bermuda and penetrated the air defenses of the northeastern continental U.S. During part of this penetration all bombers were painted simultaneously by the "Millstone" Lincoln Laboratory radar. Mission 35-mm "time-lapse" photos of ground radar scopes were used to determine precisely bomber positions compared with teller's log plots, and were also compared to navigators' logs. Incidentally, B-47s were positively identified from the commercial traffic with high confidence by a simple method:

Since all B-47s were dispensing chaff during this penetration, a 35-mm Movieola film editor with forward and reverse treadle control was used to run the 35 mm film strip forward or backward to see if the B-47 swallowed chaff. If so, it was known to be one of the B-47s in the formation.

This large mission with over 2,000 tellers' log plots was analyzed manually by Culling and also programmed for analysis by the SAC Hq IBM 7090 computer, so that the computer program could be checked by his manual analysis results. The results proved that he had derived a much better evaluation method for tactics and ECM than was previously available. This was our contribution toward more objective analysis of ECM data. However, a short time later this ECM wing was transferred to Eighth Air Force, and we went on to other things.

I mentioned to Maj. Gen. McConnell that we thought there was a real need to speed up alert response time because of the increasing Soviet ICBM threat. We also saw that the planned permanent and hardened alert facilities that SAC wanted to build at its bases to house alert crews near their aircraft would not be available for perhaps five to seven years because of the lengthy process required to get permanent construction programmed, budgeted, and approved by Congress. I told him that we believed some mobile homes with extra soundproofing could be obtained quickly and placed in the alert area. Also we had observed that under our present procedures the first eight or ten minutes generated virtually no takeoff traffic because during this time the aircrews located elsewhere on the base were picking up their mission folders and then traveling to their aircraft. Then there was a queueing problem so that only a few could taxi before others which were waiting, in order to preserve a reasonable time between takeoffs.

We proposed an "alert" part of the alert force which could be only four or six aircrews, who would need to stay near the aircraft so as to fill up some of this dead time at the start of the takeoff sequence. General McConnell listened and thought it was worth considering. A few minutes later I received a call from Colonel Thompson, Director of Operations, who said, "Let's go look at some trailers." We did, and in less than two weeks we had a test facility in operation by the 301st Medium Bomb Wing at Barksdale AFB.

A few weeks later I was invited to attend a semiannual war-planning conference at Hq SAC as a member of the Second Air Force team of planners. While concluding this conference, General Power stated his need to get better bomber alert response time because of the emerging Soviet ICBM threat. General McConnell then said, "We've got what you need down at Barksdale under test now. You're invited to come down and see it." General Power came to visit the facility soon afterward, and then ordered every commander in SAC to visit the facility at Barksdale and see how the Second Air Force was doing it.

Some months later I was given a surprise invitation to a special staff meeting, where General McConnell awarded me the Meritorious Civilian Service Medal. Naturally I was pleased that we had contributed in some small way to our improved national defense posture.

Such projects and numerous unmentioned others helped Operations Analysis find a place in the Second Air Force staff, and the acceptance we had earned was heartwarming. We had one firm policy that we followed. We tried to make sure that all staff personnel we worked with understood that we were there "to help them look good in front of their superiors," and not "to put them down." However, we tried to be a "good conscience" in that we would always argue for what we thought was best for the nation, SAC, and Second Air Force — in that order.

Toward the end of 1959 I received an unsolicited offer to be considered for a supergrade PL-313 appointment to become a scientific advisor for Systems to the commander of the Aerospace Technical Intelligence Center (now Foreign Technology Division) at Wright-Patterson AFB, Dayton, Ohio. It was with great regret that I left Second Air Force and SAC. I believed then, as I do now, that the four years spent as Chief, Operations Analysis, at Hq Second Air Force was the most challenging, the most responsible, and the most satisfying of all the positions I held in the 32 years that I worked for the Department of Defense.

Hq Eighth Air Force (Reported by Tom Chappelle)

On 7 June 1946 Hq Eighth Air Force was relieved from assignment to USAF, Pacific, moved (less personnel and equipment) to MacDill Field, Florida, and was assigned to SAC. On 1 August 1946 Hq Eighth Air Force was attached for administrative purposes to Hq Fifteenth Air Force.

Hq Eighth Air Force moved from MacDill to Fort Worth Army Air Field, Texas. Approximately half of the fully equipped combat units of the Fifteenth were transferred to the Eighth on 1 November 1946, but they remained under the Fifteenth's administrative control until 19 November 1946, at which time Hq Eighth Air Force became operational.

In the realignment on 1 April 1950 the Eighth was assigned both bomber and reconnaissance units in the Central portion of the United States.

Effective 13 June 1955 the Eighth was moved to Westover AFB, Massachusetts, and assumed control of units in Northeast and Central U.S.

In the realignment of 1 January 1959 several bases and units were exchanged between the Second and the Eighth. This placed the Eighth in command of forces in Eastern United States.

On 1 July 1963 a realignment placed the Eighth into an operational ICBM environment in the Midwestern and Rocky Mountains regions. The Eighth acquired from the Fifteenth a Titan I Wing and an Atlas Wing at F. E. Warren AFB, Wyoming, and from the Second, Titan IIs at McConnell AFB, Kansas. Also Whiteman AFB, Missouri, with its embryonic Minuteman Wing.

On 31 March 1970 Hq Eighth Air Force ceased operations, leaving SAC with two numbered Air Forces.

Because of its illustrious history, primarily in World War II in Europe, the Eighth was not inactivated. The numerical designation was transferred to Guam, absorbing the personnel and equipment of Hq 3rd Air Division. Hq 3rd Air Division at Andersen was inactivated.

At the outset there were only three SAC field units, one Operations Analysis Office at each numbered Air Force. The office at the 1st Missile Division came later. Initial negotiations concerned discussions about the critical mass required for an office to be truly effective, and the result was the decision that the offices should consist of three professionals and a secretary. In true Operations Analysis fashion, a numbered Air Force would have an office only if it wanted it and made the necessary resources for it available. All three did. With those and some other decisions made, I was sent to each of the three numbered Air Force Hqs to discuss procedures and requirements for establishing and manning the offices. I met with fine cooperation and a considerable amount of enthusiasm.

Meanwhile, back at headquarters, it was becoming apparent that providing chiefs for the offices at Second Air Force and Fifteenth Air Force was not going to be a problem, but that Eighth Air Force, which was scheduled to move in early 1955 from Carswell AFB, Texas, to Westover AFB, Massachusetts, was another matter. Originally I had no thought of leaving Hq SAC. Besides, if I had, my natural inclination would have been toward the mild Southern California climate.

In the long run, several factors influenced me to consider setting up one of the offices and to select the Eighth Air Force office as the one to set up. The main factors were (1) the job needed to be done, (2) I was convinced that experience in the field would be valuable training, and (3) my greatest satisfaction had always come from establishing new endeavors and carrying them to the point where they were operating smoothly and effectively. Another factor was one that made it easier to consider the Eighth AF post. I had been for some time a member of the Advon that was scheduled to deploy to the United Kingdom in the event of an emergency. That Advon consisted principally of Eighth AF personnel and was commanded by the Commander, Eighth Air Force.

Still another factor was that Commander, Eighth AF, was Maj. Gen. J. B. Montgomery, with whom I had earlier worked closely at Hq SAC, and for whom I had a great deal of respect. However, after I committed myself and was preparing for the move, Monty resigned from the Air Force. Fortunately what could have been an uncomfortable situation worked out very well. Cam Sweeney moved from Fifteenth AF to Eighth AF when it moved to Westover AFB. I could not have asked for a better boss. More about that later.

A small advance party from Carswell and I were the first arrivals at Westover. There was plenty to keep us occupied. Westover had for some time been the East Coast terminus for overseas flights for the Military Air Transport Service (MATS), whose former functions were now transferred to McGuire AFB in New Jersey. Some of the Westover facilities reflected the distinctive nature of the functions the base had been serving. For example, the officers' club served many more transients than permanent party officers. Modifications to fit Eighth AF needs were extensive and were accomplished in the self-help manner. We in the Operations Analysis Office pitched in and helped, too.

The analysts in the office until Bruce Stafford replaced me were Al Johnson and Frank Heilenday. Al came from the Naval Ordnance Laboratory, had an M.S. in physics, and had the distinction of having nine children, including twins. He and his wife, Harriet, just liked children. They were both Protestants, products of Wheaton College. When Al left SAC he went to IBM's Federal Systems Division in Owego, New York, and has been there ever since.

Frank came from Convair, San Diego. He is still in SAC, currently in the office at Hq SAC that was formerly Operations Analysis and in 1971 was changed to the Science and Research Office. Frank is a New Jersey native son and has an M.S. in aeronautical engineering from MIT. I remember that he took work toward a doctorate.

General Sweeney and his staff did everything they could to make us feel like members of the family. I was granted the privileges of a director, including that of being able to get airplanes for transportation. There was one difference with regard to the airplanes, though. The other directors were all pilots. I had to be furnished a pilot. When I flew in a T-33, as I frequently did, the pilot had to be an instructor pilot.

The number of officer's houses was limited, but I was offered one of the houses. I declined it, but was gratified by the offer. I participated in the senior staff meetings, in important headquarters meetings such as the periodic briefings by the 8AF Wing Commanders, and in the 8AF Commanders Conferences. My wife and I were regularly invited to the official receptions and parties, and at every one of them General Sweeney made a special point of coming over to chat with Jeannie and making her feel fully included in the occasion. He was a real gentleman and, obviously, very thoughtful. I also knew I was included in the inner circle when he made a habit of offering me a cigar when I had a session with him in his office. He was lucky; I wasn't a cigar smoker.

In addition to tasks General Sweeney assigned us, we worked principally with Operations, War Planning, and Intelligence. We also provided a considerable amount of day-to-day assistance to them and to other parts of the headquarters on a consulting basis. Buzz Humfeld was director of Operations, John Henshke was director of War Planning, and Rocky Triantafellu was director of Intelligence — all colonels at that time. I was still a member of the U.K. Advon, and we had periodic tests of our ability to assemble and board the airplane (Sweeney's C-97) on short notice, including calls at home in the wee hours of the morning.

I also was a member of the Eighth AF Planning Committee and of the alternate headquarters planning group. Because of my nuclear weapon background I served as one principal contact for the nuclear officers in the field units.

The first major project we undertook was to determine the safest location for the alternate headquarters with relation to probable fallout from an initial nuclear attack on the United States. Later, after a general area had been selected, we helped to select a specific location that offered good protection because of the lay of the local terrain.

It was during my two years at Westover that the B-52 landing gear problem reared its ugly head. We did some small analyses related to that problem.

A much larger project concerned refueling rendezvous. It was very much an Operations Analysis project. In addition to analyzing the refueling procedure, we mounted cameras in airplanes and flew on many missions recording scope pictures that showed that we had beacon problems and other difficulties that hadn't previously been fully recognized.

Another major project concerned the noise generated by the operation of B-52s out of Westover. It resulted not only in the identification of physical arrangements and operating procedures that would help to alleviate the problem, but also in recommendations that were adopted relating to public relations efforts with surrounding communities, and to providing advice to them on construction and siting of residential and commercial structures.

Late in my two years at Westover the senior officers responsible for communications in the Eighth AF operating area came to the headquarters with a briefing describing changes and additions needed in the network. General Sweeney thought that the briefing was not ready to be carried to Washington and turned in his chair to say to Buzz Humfeld and me something like, "Buster and Tom, in coordination with these gentlemen, revise and recast the briefing to present the message we want to convey to Headquarters USAF."

We did so and accompanied General Sweeney to Washington and the session with Headquarters USAF. Buzz did the presenting, and the briefing went over well. Years later when Buzz had retired from the Air Force and was in industry, we chanced to see each other in Washington. In introducing me to another representative of his company, he presented me as the person largely responsible for his promotion into the rank of general officer and mentioned specifically the success of our cooperation on that communications requirements briefing. Buzz overstated the case, because he was a competent officer in his own right. However, he and I did work well together.

Those are the highlights of my time at Westover. I returned to Omaha at Rod Smith's request; I had gone to Westover with the understanding that I would stay long enough to get the office set up, manned, and operating effectively, and that I would return to Omaha when the need for me there was great enough after I had accomplished those goals.

Hq Eighth Air Force (Reported by Bruce Stafford)

Heading the three-man Operations Analysis Office at Eighth Air Force, Westover AFB, was a complete change from being an analyst at Hq SAC — and a great experience. Not only were we brought much closer to the "troops" in the operational units and all of their "real-life" problems, but we were completely absorbed into a somewhat smaller (than at SAC) headquarters organization led by Maj. Gen. W. C. Sweeney, one of the most able military men I have ever known. He believed in us and our function, and saw to it that we were part of everything that went on in his command.

Shortly after my becoming Chief, Operations Analysis (in early 1957), the alternate Hq SAC control room that was being chiseled out of a nearby mountain was completed to the stage where it was usable during alerts, CPX's, and large-scale exercises. Al Johnson, Frank Heilenday, and I spent a great deal of time designing and operating various control devices and displays in the underground command post, trying always to improve the Boss's understanding of the status of his operational forces (tankers and bombers at that time), and his capability to direct them. We had our own station on the "command balcony" and pulled around-the-clock duty right along with the military, during these alerts and exercises.

General Sweeney was a farsighted man, interested in the future and all the possibilities that go with it. I found myself plunged into space activities, at least to the point of being able to explain to the "old man" what Sputnik was, and what it would mean to SAC, in particular, and to the United States in general. This was one of the many times when I leaned heavily on the SAC Operations Analysis Office for support. The backup they provided in the form of memoranda, briefings, and temporary duty by SAC analysts was invaluable, and allowed us in our small office to function much more effectively.

In many cases we were able to get new information to Eighth Air Force before it could arrive by any other means, for example technical data for new equipment, test-firing results of nuclear devices, performance data of missiles scheduled to be assigned to Eighth AF, developments in space technology, and so forth. This line of communication worked very well in the other direction as well. I recall that we were able to collect a considerable number of scores on bombing runs from our Eighth AF units for SAC Operations Analysis, as well as a great deal of numerical data from our Eighth AF bomber crews as they performed "breakaway" maneuvers designed in Mart Schwartz's shop.

The list of projects we worked on in the two years I was Chief, Operations Analysis, Eighth Air Force, is long and varied; a complete accounting may never be possible; I can remember only a fraction of the total. However, some remain rather vivid memories, principally because Eighth AF was a northern-based command and, therefore, had to deal directly wth the elements, particularly the ice and snow of winter.

One project required us to develop a measurement technique to determine "slickness" of an icy or wet runway, so that pilots could be provided a reliable number to be used in planning, and for takeoffs and landings. We were able to borrow and modify both equipment and technique from the Danes, who had considerable operational experience at their Greenland bases. Both Al and Frank collected data firsthand under various slippery conditions at Eighth AF bases, riding the runways in a pickup truck while measuring stopping distances, coefficients of friction, accelerations. I believe that the numerical tables we developed and passed to the SAC units were the first successful attempts at improving on the "eyeball" guess of runway condition.

From my own selfish viewpoint, the most satisfying and most important accomplishment to come from my tour in Eighth AF was the development of a close working and personal relationship with General Sweeney, his staff, and his commanders. Whether it was at a staff meeting, in a review of a unit's war plan assignment, on a business trip with the Boss in his aircraft, in the command post, or at a Commanders' Conference, we analysts were expected to participate fully — which we did wholeheartedly. I discovered later, when I returned to Omaha, that the mutual respect and friendships developed during those two short years enhanced considerably the working relationships with the SAC staff, especially with those officers who had been in Eighth AF.

* * *

Returning to Hq SAC in early 1959 as a team chief, then deputy, and finally chief in 1963 gave me an ever-changing and more complete view of the workings of SAC (undergoing considerable change as well), and our relationships in Operations Analysis with the military staff. My experiences during those years, and for some time after, I cherish especially, principally because of the dynamic leadership of General Tom Power. With the advent of the ICBM and other sophisticated systems, our workload in Operations Analysis seemed to become suddenly more complex and broader in nature.

In my team, Operational Capability, we were trying to optimize Minuteman basing, developing teaching-machine methodology and applying it to the learning of "must know" rules used in command posts. This effort included developing tactics to improve effectiveness of our airborne alert sorties then flying; studying and testing a mobile (railroad trains) ICBM system to determine feasibility and practicality; applying the scientific method to the design of missile firing tests, and analyzing results to insure that SAC planners got the best and most information possible from these expensive "destructive" experiments; and on and on.

In the case of ICBM firing tests (on both Atlantic and Pacific ranges), it seemed to me that we spent an inordinate amount of time explaining proper experimentation techniques to military officers and others not schooled in the subject. However, I believe it was worth the effort, and served to demonstrate again one of the functions of Operations Analysis: do the initial job as required and continue to do it until others (staff agencies) are "brought up to speed" and can take over prime responsibility. It was only a matter of time until terms such as "confidence limits," "sequential testing," "confidence levels," and "sampling" became commonplace in the vocabularies of almost everyone on the SAC staff. On one occasion, I remember that I made a quick departure from a senior staff meeting, at General Power's request, to find out what level of confidence was claimed in a typical national poll. Fortunately, I reached Dr. George Gallup by telephone, got a direct answer from him and reported back to the staff meeting before it adjourned. I'm sure this contributed to everyone's understanding of estimating by means of sampling techniques, for it showed them a consistency exists even though sample sizes and fields of endeavor differ greatly.

Not only was General Power a dynamic leader of military men and women, he was an articulate advocate of strong strategic forces. He could persuade audiences that his viewpoint was rational and correct — a real spellbinder. He always looked to the future, while making certain that present-day operations were under control — whether it be listening to Dr. Percy Carr (Iowa State University Operations Analysis Standby Unit) explaining Einstein's Theory of Relativity and related space topics, or directing me to contact the top nuclear physics experts in the country in order to collect their thoughts and opinions related to new developments possible in nuclear weapons and devices.

After General Power, there was a succession of relatively short-term CINCSACs. Perhaps this reflected the times, since the nation was becoming more and more committed to military operations in Southeast Asia. And SAC changed; the development of new strategic weapon systems virtually stopped as more and more manpower, material, and money from SAC resources were siphoned away to Vietnam and all that went with it. Our office took on several projects directly related to the Southeast Asia operation, while continuing the ongoing projects necessary to maintain strategic nuclear deterrence.

At the outset of SAC's participation, in a conversation with Maj. Gen. Bill Crumm one day I agreed to provide an analyst to support him at the Third Air Division at Guam, at least until he got things running smoothly. This we did; several analysts served for some months on a rotational basis, providing direct assistance to General Crumm and members of his staff (principally Operations, Intelligence, and Material) and serving as a link to our office for additional support.

Throughout the Vietnam conflict Glen Hadsell, in our office, worked tirelessly to improve SAC bombing capability with "iron" bombs, adapting to changing conditions and requirements by modifying equipment and procedures, by using electronic transponders (beacons) in novel ways. He traveled to the field again and again to work directly with the aircraft crews, then back to the staff, and so on, until the improved capabilities were included in combat aircraft. I think "Mr. Beacon" was one of the unsung heroes in that whole operation in Southeast Asia.

Back in Omaha we were trying to evaluate the B-52 raids on South Vietnam — to determine the "bombing effectiveness" which seemed to be an obsession with nearly everyone in the Pentagon. This was a seemingly impossible job, since the targets were generally foot-soldiers moving under jungle canopies, and little or no direct bomb damage assessment (DBDA) was possible. Things were complicated further because there was a dearth of technical information related to analyzing World War II-type train bombing, requiring us to revert to basic mathematics and to rederive many relationships.

Probably my most sobering experience during the Vietnam era came when I was called suddenly by General John C. Meyer, then CINCSAC, to participate directly with him and a few senior staff officers in a back room of the command post, during execution of LINE BACKER II. The quiet professional way this group acted is something I shall never forget. I couldn't help thinking of Zimmie while we wrestled with the alternatives of tactics of execution, realizing that he and others on SAC crews involved were facing a completely different situation from that over the jungles farther south. Later, I was most pleased that Zimmie (all 6'4" or so) stopped into the office to say hello, and allowed us to debrief him a bit.

Fifteenth Air Force

Effective 21 March 1946 Fifteenth Air Force was assigned to SAC and the headquarters were activated at Colorado Springs, Colorado. Approximately half of the Fifteenth units were transferred to the Eighth Air Force.

In 1947 General Clements McMullen, Deputy Commander SAC, relieved several units from command jurisdiction of the Fifteenth and assigned them directly to Hq SAC. The Fifteenth was left as strictly a bomber command.

On 7 November 1949 Hq Fifteenth began operations at March Air Force Base, California, having moved there from Ent Air Force Base, Colorado.

Hq Fifteenth in California controlled units at MacDill AFB, Florida. The 1 April 1950 realignment assigned bases and units only in the Western area of the United States to the Fifteenth. Realignment on 1 April 1955 made the Fifteenth responsible for bases in the Southwest and West. In a realignment on 1 January 1959, the Fifteenth's area of responsibility remained in the Western section.

On 15 January 1959 the 703d Strategic Wing (ICBM - Titan) at Lowrey AFB, Colorado, and the 706th Wing (Atlas) at Francis E. Warren AFB, Wyoming, were transferred to the Fifteenth. On 11 July 1963 the Eighth AF acquired from the Fifteenth the missile wings at Lowrey and F. E. Warren.

Realignment in 1970, when the Eighth ceased operations, made the Fifteenth responsible for all combat ICBM units, the entire strategic reconnaissance force, and a few B-52 and KC-135 units.

Ed Winger reported to Hq SAC in November 1950. He has written:

I was assigned to Stan Lawwill, who was head of the Atomic Energy Division. That was the best thing that ever happened to me. Stan was a wonderful boss. Later on, I became head of the Atomic Energy Division under Rod Smith, when you became Chief Scientist, SAC, took Stan as your deputy, and made Rod Chief, Operations Analysis.

General LeMay was polite, courteous, and his comments were short and to the point. He was a gentleman.

I was assigned to Commanding General, SAC X-Ray, which was manned by Fifteenth AF personnel and had responsibilities in the Pacific in the event of an emergency. I remembered during the Korean conflict you called me out of bed one Saturday morning, and you sent me to Japan for an indefinite period that afternoon, with nothing but dirty clothes and no money.

I checked in with General Rosie O'Donnell. I could do whatever fuzing job was required, because I had the information, and you sent me items to keep me current. Rosie O'Donnell told me that the Army asked that I be assigned to them. My God, maybe they couldn't fuze their own artillery shell. I could. Rosie said, "NO!" so that was the last I heard of that.

General Sweeney, after he took command of the Fifteenth AF, said he wanted Operations Analysis, Hq SAC, on his squawk box. That didn't work out, so I went to March AFB as Chief, Operations Analysis, of the Fifteenth. General Sweeney was most cooperative. He set us up real well, and treated us real well. He regarded us as his personal staff. We undertook some projects for him that the staff knew nothing about.

Author's note: I visited Cam Sweeney at Hq Fifteenth AF the day after he assumed command. He told me, "Zim, within two hours after the Change of Command ceremonies, I was sitting in this chair when General LeMay phoned me and really chewed me out about a personnel problem in the Fifteenth. I obviously could not have been involved." Later I found an occasion to mention this to General LeMay. His reply: "Zim, that is one of the hazards of being a commander."

General Sweeney went as Commander, Eighth Air Force, and General Archie Old became Commander of the Fifteenth. In the beginning it seemed as if he treated us with benign neglect. He accepted us and was polite enough, but we sometimes wondered about our usefulness.

Then you made a trip to the Fifteenth. In anticipation of your visit we sent to General Old a summary of all the projects we had worked on. He was impressed, because from that time on we were treated as regular members of the staff.

* * *

Rick Camp wrote:

When Ed Winger and I arrived at Hq Fifteenth, General Sweeney was Commander. His first assignment to us shook us to our boots. It was not that the problem was so special; it was the manner in which he asked us to work on it. It was a communications problem, and as I recall he suggested something like, "What I would like you to do is to snoop around and find out what we are doing wrong in the communications business."

Knowing General Sweeney as I know you did, Zim, surely you would have understood his meaning better than we did at that time. However, it was clear that he wanted some effort to look closely at the communications problems of the Fifteenth. Ed and I decided that "snoop around" did not mean covertly. We went directly to the colonel responsible for communications and offered our assistance, because he certainly was as interested as we were to find solutions to General Sweeney's concerns.

It took a fair amount of tact and some of Ed's good management to get the colonel on our side — more properly, get us on the colonel's side — and get us working toward a solution. I don't recall how far we got toward a solution, as about a month after we were assigned the problem General Sweeney got orders to the Eighth Air Force, and was replaced by General Old.

I have the greatest respect for General Sweeney's approach to management, which was very different from General Old's. General Sweeney got very much involved in the details of his command, whereas General Old managed much more by exception.

I remember Russ Dougherty being one of our contacts and one of those with whom we worked closely. Several of our early studies were involved with the B-47s, which were just then becoming operational. One study was the effect of high altitude and the possibility of the loss of pressurization, which greatly worried the operators. At that time I was reminded again of the need for a team approach. We became heavily involved with tracking down the medical problems of individuals at low oxygen levels and under conditions of low pressure.

We had many exercises with NORAD, and we were responsible for their analysis. They were on a once-a-month basis.

We were heavily involved in support of the War Plans staff. Every year we would gather in the basement of Hq. SAC to plan and analyze the Emergency War Order.

1st Missile Division

Located at Cooke Air Force Base (renamed Vandenberg), California, the 1st Missile Division on 1 January 1958 was transferred from the Air Research and Development Command to SAC.

Hq. 1st Missile Division was redesignated Hq. 1st Strategic Aerospace Division on 21 July 1961.

Dr. Fielding McGehee was Chief of the Operations Analysis Office at 1st Missile Division.

SAC MIKE (Reported by Rick Camp)

[On 1 January 1958 the Office of Assistant CINC-SAC (SAC MIKE) was organized at the Ballistic Missile Division of the Air Force Systems Command at Inglewood, California. Designed to serve as an extension of Hq SAC, this office was responsible for working closely with the Ballistic Missile Division and providing SAC with information on ballistic missiles, the latest techniques and related programs. SAC MIKE was discontinued on 1 July 1961. Rick Camp from Operations Analysis Fifteenth Air Force was asked to head an office composed of three civilian analysts.]

Our responsibility was to assist Colonel (later General) Large in his purpose of introducing SAC to the ballistic missile program. The development of the ballistic missiles had been the responsibility of the Air Force Systems Command under General Schriever. In the early days there had been consideration that operation of the missiles might come under some other command than SAC. The decison was made, however, that SAC should become involved, and that ultimately the missiles would be under the operational control of SAC.

To expedite development of SAC specifically in the operational use of ballistic missiles, SAC MIKE was conceived as a part of SAC Hq, and located at the Ballistic Missiles Division. It was the plan that SAC MIKE would become involved in the development and operational planning in the use of the missile. Sometime in the future SAC MIKE would be phased out and the personnel and knowledge gained would be phased into Hq SAC at Omaha.

Thus it was that when I first reported to Colonel Large his first words to me were that my job was to help him work himself out of a job. He visualized this as about a three-year operation. In fact, SAC MIKE was disbanded after about two and a half years.

The first job I faced was staffing the office. I was assisted in this by the Operations Analysis Office at Hq SAC. I'm reminded of the importance in my mind at that time, drawn from my association with Zim, of staffing the office for a team approach to the problems. So it was that I thought in terms of a mathematician, a physicist, and an aeronautical engineer. It has been of interest to me in later years to observe that the team approach which Zim emphasized early, for both office and project manning, is now written as one of the characteristics of operations research in textbooks on the subject of quantitative approach to management.

Our earliest studies were involved with the siting of both Atlas and Minuteman missiles. Working with the planning staff of AFSC (Colonel Lulejian, Lt. Col. Herb Hoover, and others) we looked in considerable detail at the problems associated with where the missiles might be situated. Specifically at SAC MIKE direction, the distances between sites were determined to prevent destruction of more than one missile by an incoming missile. Many of our studies were accomplished working with the staff at AFSC.

I'm reminded of some of the work of Dr. Alden Cleaves on my SAC MIKE staff. I think that Alden's conceptual understanding of SAC problems, and his capability to view those problems from the broadest viewpoint, were his greatest contribution during my close association with him.

During those early days, morale was high at SAC MIKE, and the learning curve was steep. For both civilians and military alike, there was much to be learned about these new things called ballistic missiles, and much to be learned about space and what went on in the space environment. I remember the enthusiasm with which people like Col. John Fahringer, a fighter pilot from Hq SAC, undertook his job in the space business, and how he became a devotee of the concept of SAC in space. Our early efforts were in support of John's concerns with how better to use space in support of the SAC mission.

The most interesting, and the largest in terms of manpower and responsibility, of all my studies during my stay at SAC MIKE was that of the mobile Minuteman system. I headed the study of the use of the Minuteman on railroads, at that time a favored operational concept of both General Schriever and General Power. Working with a 30- to 40-man team, including support from several major railroads, we were to evaluate the feasibility of the Minuteman system being placed on rail cars, in a mobile operational mode. We evaluated the technical and operational feasibility, and cost and cost-effectiveness.

Under technical feasibility we looked at the requirement for modification to the missile, as well as to the warhead, to make it possible to handle and use effectively a missile originally designed for a fixed-base system, operating on the roadbeds of the rail system in this country. We developed an operational concept for use and support of the mobile system. Many of the ideas of mobility for missiles have continued to the present day with the MX system.

We made detailed assessment of the cost of the system, including the development and modification costs to the missile and the rail system. The operational cost showed that although costs would be greater than for the fixed system, they were not prohibitive. What started out in the study group as an excitement and enthusiasm for a new system, gradually developed, as the study progressed, into concern for its cost-effectiveness, as opposed to its feasibility.

When the study was completed and the presentation was being prepared for General Power and General

Schriever, there was considerable question as to how the presentation could be made to give an honest view of our beliefs on the proposed system which originally had generated enthusiasm both at SAC and AFSC Hq. It was agreed that if we presented our findings on cost-effectiveness early in the briefing there was a good chance that we would never finish all the aspects of technical and operational feasibility, which we thought were important to present.

The briefing was designed to present technical feasibility first. Conclusion: Yes. Operationally feasible: Yes. A workable operational concept had been developed and could be supported by the maintenance staff at SAC Hq. Economically feasible: Yes, although slightly more expensive. Finally in the last four charts we presented comparisons in terms of the number of warheads on target. The conclusion was clear that we did not recommend proceeding further with the proposed system.

I will never forget General Power's response. Throughout the briefing he had been complimentary and obviously interested in the results of the study. His remark at the end of the briefing will live with me. "I liked everything about your briefing, except the last two charts." He directed that we work with his staff to investigate additional items of concern to him, which we did during the next few weeks.

A contribution to SAC and the nation, the study group was instrumental in turning off a system which we believed then, and do now, would be a costly mistake. The primary difficulties with that system were knowing its location and the shortness of the intelligence cycle.

My 18 months at SAC MIKE were certainly exciting and productive. But all of my 12 years in SAC were exciting, and among the most productive of my career. I feel a deep sense of gratitude and a debt to Zim for getting me involved in a career that I can look back on with some pride and some sense of contribution to the nation.

Chapter 9

With General LeMay at Hq USAF

Shortly after General LeMay went to Hq United States Air Force as Vice Chief of Staff, USAF, he asked me to join him. I refused. Later he asked me the second time, and I refused. Still later he phoned General Power, then CINCSAC, and asked if he could speak to me once again.

This time, in the Pentagon, he told me that Operations Analysis had been elevated to report directly to the Vice Chief of Staff. The position of Chief, Operations Analysis, had been allocated the highest Civil Service rating, and a deputy had been authorized. (My predecessor, Roy Brothers, did not have a deputy.) Would I reconsider and accept the position of Chief, Operations Analysis, USAF?

Before I answered he said he wanted to warn me that I would never like it there. It was practically impossible to get anything accomplished. At that time I didn't actually know what my response would be. But as he walked me to his door he said, ''Phone me and tell me when you want to report.''

* * *

General Power, then CINCSAC, had a reputation for chewing out people, which he did with expertise. I was embarrassed when he really chewed out a major general in my presence, but I had so far escaped his wrath.

After I got back from my talk with General LeMay, the squawk box would frequently sound from the Commander. General Power would say, ''Zim, have you considered this?'' ''Have you thought of that?'' So when I went to General Power to tell him that I had decided to go to Washington, he royally lit into me. He racked me up one side and down the other. Remember, he was an expert. He always thought I had more sense than to go to Washington, etc., etc.

I learned the next day that after I left his office, he told his Chief of Staff, Maj. Gen. Eddie Broadhurst, to write me up for the highest award the Air Force can give a civilian — an award which I received from General Power before I left SAC.

Because of the school year, it was several months before I would report to Washington. I was not ostracized. During the remaining time, I was kept in the ''inner circle'' and treated as if I would forever be with SAC.

* * *

Soon after I reported to Hq USAF, one morning I checked General LeMay's schedule with a colonel, in

Receiving an award from General LeMay, Vice Chief of Staff, USAF, 1960.

With Secretary of the Air Force Eugene Zuckert, 1962.

Giving award to Ken More, Hq USAF, 1962.

Giving award to Clayton Thomas, Hq USAF, 1962.

Giving award to Walter Deemer, Hq USAF, 1962.

his outer office. I had not known the colonel before. I selected a time and asked him to put me down for it. He asked the subject. I told him, and he said, "Do you think you ought to bother him with that?"

I replied, "Colonel, the first time you see me thrown

With Dr. Martin D. Schwartz, Hq USAF, 1962.

Receiving an Honorary ScD. from Dr. Louis Perry, President of Whitman College, 1962.

out of that office, ask me the same question." I was never thrown out.

My move to Hq USAF was not a promotion as far as Civil Service ratings were concerned. At SAC I received the highest salary of a PL-313 position, which was the same rating I assumed at Hq USAF. This was equivalent to a GS-18, but for research types, primarily.

* * *

Six months after General LeMay went to Hq USAF as Vice Chief of Staff, he and his wife, Helen, came back to Omaha for a few days: he was to be briefed by SAC, and they would attend the Aksarben activities. (Aksarben is Nebraska spelled backwards, which may seem corny, but they thought of it in 1895.) This was the elite social event of the year for Omaha, and one of the most unique and impressive in the U.S.

Modie and I were always guests at the yearly coronation of the King and the Princess of Aksarben, and at some events afterwards. The LeMays were at an exclusive party at one of Omaha's hotels after the coronation.

When we got home after the parties I received word that my father had gone into the hospital in Dayton, Ohio, in a critical condition. I tried to get airline reservations, but none were available for three days. I knew that General LeMay was leaving the next morning for Washington, so I went to the hotel where I knew he was. I was not then in formal dress, and I was stopped trying to go into the party room. I told the man who stopped me I wanted to see General LeMay. He said General LeMay was unavailable. I told him to tell General LeMay that Zim wanted to see him, and let him be the judge as to his availability. The general got the message and immediately came out to the hallway where I was waiting.

I told him my problem, and asked if he could drop me off at Dayton on his way to Washington. He said, "Certainly. Just be there a few minutes before take-off time."

General LeMay was flying a C-131, and as always during this period, he was up front all the time. The only passengers were Helen LeMay, the aide, and me. I knew Helen well. We talked, just the two of us, at length. She told me this story: While at SAC General LeMay came home one night and asked if she and Janie had any plans for that night.

"No."

"Good. We have a family problem to discuss."

After dinner General LeMay said, "They have offered me the job of Vice Chief of Staff in Washington. If I don't accept I will stay here for a year, until I retire.

"Janie, you were never particularly fond of military quarters. You both know that I could get a position that would pay much more than I can get with the Air Force. If I retired we could have a fine home, I could buy you a new sports car every year. I could also buy your mother the mink coat she's always wanted, but I couldn't afford. What do we do?"

They discussed the pros and cons, and then decided he must go to Washington. He said, "I am glad it came out that way. Although I don't know if I can make it in Washington, I know I'd never be completely content if I didn't try."

That is the General LeMay the press and the public do not know.

* * *

According to General LeMay, Helen "was up to her ears in every sort of Air Force Wives' activities" and other women's activities connected with the armed forces. A committee of officers' wives from the joint services renovated an old house in Washington for transient personnel and later added a 55-bed annex. This hostel filled a real need.

At Omaha Helen was quite active in the Officers' Wives Club of Hq SAC. Civilians of a given rank, or higher, were eligible for associate membership in the Officers' Club. This meant they had all the privileges of the club except voting. Membership was not automatic. They had to be voted in. Wives of those civilians who were members of the Officers' Club were eligible to be associate members of the Wives Club. Again it was not automatic. Helen routinely asked me to approve the wife of a civilian before the Wives Club would vote on her.

* * *

I was returning from Europe on a KC-97 with General LeMay. We remained overnight at the air base at Reykjavik, Iceland. The next morning we were aboard the aircraft at takeoff time, with General LeMay strapped into the left seat in the crew compartment. (In early days General LeMay flew each aircraft each minute we were in flight. In later years, he would come back and spend some time with those of us in the rear of the aircraft. But he still made each takeoff and landing.)

All were aboard except two colonels. A jeep was racing toward the aircraft with Col. H. R. Sullivan and Col. A. J. Russell. As they got to the aircraft Colonel Russell realized that in his haste he had left a $300 watch he had purchased on the dresser in the VOQ. He took the bull by the horns and went up front to ask General LeMay if he could get his watch. General LeMay grunted.

Russell was not gone long. After he got back on the aircraft General LeMay unstrapped himself, walked back, looked at Colonel Russell, waited a moment, and then asked, "Is it O.K. to take off now?"

* * *

Once when returning to Omaha from a staff visit to Europe, Stan Lawwill was sitting in the rear of one of the SAC VC-97s talking with a group of SAC officers. He commented that he had been mightily impressed with the shrewdness and logical clarity of the response General LeMay had made to a penetrating question asked at one of their stops in Europe. Col. Dave Jones, then General LeMay's aide and later Chief of Staff, USAF, and Chairman of the Joint Chiefs of Staff, had an immediate explanation. He said, "Doc, while you and I and these other guys are sitting back here talking about women, the Old Man is up there in the crew compartment thinking abut SAC."

* * *

While en route to Europe in a SAC VC-97, Stan Lawwill was having lunch with General LeMay and others. He commented that he had recently attended a party in Capt. Alan Birdsall's quarters, which were in the Wherry Housing just built near Offutt AFB. He was aghast at how thin the walls seemed to be; whenever there was a moment of quiet in Alan's quarters, they could clearly hear the conversation next door. Stan said, "That's community living."

General LeMay remarked that his first on-base quarters after World War II were at Wright-Patterson AFB. There, enlisted men's barracks, which had been built during the war, had been converted to officers' family quarters by putting a wall across each building at the mid-point. The wall, in fact, divided the shower/latrine area which separated the two sleeping dormitories at the ends of the building. There had been serious shortages of plumbing supplies late in the war and each enlisted men's barracks at Wright Field had been built with a single valve that flushed all toilets simultaneously. When the barracks were divided into officers' family quarters, one apartment had the valve; the other did not. General LeMay said his neighbor had the valve, not he. "*That's* community living!" he said.

* * *

The SNARK (International Cruise Missile) program had problems from the start of flight testing. There were quips about "Snark-infested Tampa Bay." In a test program, one went astray and ended somewhere in Brazil. Shortly thereafter I received a copy of "The Hunting of the Snark" by Lewis Carroll from Edward Teller. He had penned in front, "To all the Hunters with best wishes, and a Merry Christmas." I still have the book.

SAC was to be the operating command for SNARK. At one of the routine status reports to the commander, the SAC Staff decided that it was time to lay out the whole program — the problems, the schedule, the decisions that had to be made, and so on. As any good status report should, this one ended with recommendations on a set of decisions CINCSAC should make.

General LeMay studied the recommendations for a while, characteristically rolling his cigar stub from one side of his mouth to the other. Then he said, "Only one of those decisions has to be made today, the second in a month, the others later. You have no idea how much smarter I'll be next month than I am now."

* * *

At a time when Stan Lawwill and I were being awarded commendations for meritorious service, General LeMay handed me a check for $300 and shook my hand. He then turned to Stan, shook his hand, and handed him a certificate, saying "That's the way it is, Lawwill. When you and your boss are being commended, he gets the bucks, you get the conversation!"

Chapter 10

After World War II: Places

Japan

Pacific Proving Grounds — for the third time I was there, but I had never been to Japan. The schedule for the next Coordination Conference in Japan fit well with the test program at the Proving Grounds. If there were no delays I could accomplish my purpose at Eniwetok and — if I could get to Japan — attend the conference and then have a ride back to Omaha on a SAC aircraft.

An aircraft was set up to take me from Eniwetok to Kwajalein, where the next morning I could catch a MATS aircraft to Tokyo. It was raining very hard with high winds. I got to Kwajalein successfully and had to get up at four in the morning to catch my aircraft to Tokyo. At that early time even the greasy spoon at the flight line was not open for breakfast. The aircraft was a C-118, and the aircraft commander was a Navy Commander. After two refused takeoffs we finally got away.

We stopped at the Navy base at Guam to refuel. I forgot to eat lunch or to get a box lunch. Indeed, I didn't know I was supposed to. When I got back aboard the crew wanted to share their box lunches with me, but I refused. The aircraft commander invited me up front for several hours. A C-118 (DC-6) was pre-jet and thus took many, many hours to grind across the Pacific. After a while I said, "There sure is a hell of a lot of water here." He replied, "Yeah, and you only see the top of it."

At 8 or 9 p.m. we landed at Tokyo International. Two officers met me. One took my passport and baggage claims, and the other took me to the VIP lounge. God knows how many hours it had been since I had eaten. Eighteen, twenty-four? In the VIP lounge there was nothing to eat. Not even a peanut. But the refrigerator was full of Lowenbrau beer, and I had four or five. This proved to be a fine way to get started on one's first night in Tokyo.

* * *

The next time I was in Tokyo I was a guest of the Japanese government, to advise them on military operations research. My host was Dr. Kaiichi Saeki, the commander of the Japanese Self Defense College (similar to our National War College). The senior military person I had contact with was General Minoru Genda, who had been one of the prime planners for the attack on Pearl Harbor. They essentially asked me if they should take the USAF Operations Analysis or a Rand route for their military operations research efforts. My reply, essentially, was that, not knowing their resources in talent or yens, I would suggest the USAF Operations Analysis route. Then later, if they could afford it, include the Rand approach. Characteristically they knew much about both.

One evening they took me to a very large restaurant. I saw no other foreigner there. We were on the sixth floor with all the restaurant below us: each table was in a private room so it required a lot of space. I was delighted to find that when we sat at the very low table there was a pit for our feet and legs. When I got back to Omaha and was telling my daughter some things about the trip, including General Genda, she said, "Oh, pals now, eh?" I could only tell her that conditions change.

My relationship with Dr. Saeki became quite close. He was our house guest once on a visit to Washington. I was able to arrange an appointment for him with Pierre Salinger, President Kennedy's press secretary. He has become an important personage in Japan. He is president of a research company which has expertise in a curious, but most interesting, combination of research and development, economics, and operations research. He is the Japanese representative on several international commissions.

* * *

On a later trip back from Tokyo I flew non-stop from Tokyo to Omaha on a KC-135. There was a nine-hour time change on a ten-hour flight. We took off from Tokyo at noon on Sunday, and landed at Offutt AFB, Omaha, at 8 a.m. on the same Sunday. As we were going in for the landing, General Power said he was going to have breakfast and then hit the sack for a couple of hours, hoping this would help him get his system back in gear. I said I was going to have a midnight snack and do the same thing. After all, it was

10 p.m. according to the time we got up by. A nine-hour time change on a ten-hour flight certainly confuses one's system.

* * *

While vacationing in the Far East in 1973 Modie and I spent a week in Tokyo. I had written to Dr. Saeki, General Genda, and Professor Imamura of the Japanese Self-Defense College. When we arrived at the New Otani Hotel, one of the finest in Tokyo, notes were awaiting me from each of them.

Though on vacation, I agreed to give a talk to their military operations analysts. Modie and I were taken to dinner by eight of them and Professor Imamura at a small, exceedingly interesting Japanese restaurant. And one evening Dr. Saeki invited us to dinner. When he arrived to pick us up in his chauffuer-driven big black car, our hotel staff could not do enough for him. We went to an elegant restaurant on the 35th floor of some building, and our table was at a picture window. It was a beautiful clear night. The lights of Tokyo below were fantastic. Dr. Saeki commented on improvements in modern Tokyo, and I told him that he mustn't forget that we helped. If we had not destroyed 75 percent of Tokyo during World War II these things probably would not have happened. He laughed, and accepted my comment in the spirit it was given.

* * *

The 1964 Summer Olympic Games were in Tokyo. That was the incentive for building freeways (downtown, too) and many fine buildings. One is the completely closed-in Olympic pool, which is used daily. The benches seat 15,000. It was pointed out that during the Olympics they seated 12,000. After the games, they repainted the bench lines designating individual sitting areas to match the Japanese anatomy. Thus they increased the seating by 3,000 without adding any more benches. The Japanese super-jet aircraft have more seats than any commercial airline in the U.S. puts in the same aircraft type.

A unique occupation exists in Tokyo. The closest English word for it is "pushers." Husky male college students are hired during rush hours to shove as many Japanese as they can into each car of the subway. They are not very gentle, and it is quite a sight.

* * *

In 1973 the yen was doing well, and some Japanese tended to be cocky in reference to the U.S. and the dollar. But today inflation and pollution have made Tokyo quite uncomfortable. We also visited some places of interest outside Tokyo and rode on the Bullet train.

I have adopted one Buddhist practice — omphaloskepsis.

Pacific Islands

I have visited the following Pacific Islands: Hawaii, Johnston, Midway, Wake, Philippines, Guam, Okinawa, Japan, Eniwetok (atoll), Kwajalein (atoll). I went to the Pacific Proving Grounds one time on a C-118 with a number of DOD dignitaries. I was the only passenger not from the Pentagon. They stopped at Offutt AFB, Omaha, and picked me up.

We remained over night at Hickam. At take-off time the next morning one passenger was missing. We waited. He came only about 15 minutes late. He was one of the sickest and most sorry-looking individuals I had ever seen. All night he had been drinking Missionary Downfalls.

Once when Modie and I were in Honolulu Wen Dwyer and Marty Schwartz were there with their wives. We had all been very close friends for a long time. Indeed, I had hired each of them into Operations Analysis at SAC — Marty from MIT and Wen from Washington University, St. Louis. This was quite a reunion. We rented an open jeep with a pink canvas top and white tassels, and together we made quite a sight. It was sometime during this visit that I ate my first orchid: they floated a small orchid in each Mai Tai, and after two or three Mai Tais, it simply seemed like a good idea to eat the orchid, so I did.

On another trip Modie and I visited Admiral Thomas H. Moorer, CINCPAC, in his office. He provided his yacht and crew to take us on a personal tour of Pearl Harbor.

I had sent Tex Alexander to be Chief of the Operations Analysis Office at PACAF at Hickam AFB, Hawaii. Sometime later he suddenly died of a heart attack. I selected Dr. Robert Hiller, a junior member of the Operations Analysis staff in Wiesbaden, Germany, to take his place. The Hillers decided to live at Kaliua, which is across the mountain from Honolulu. For some years there has been a fine car tunnel through the mountain, but at that time there was only a narrow, twisting road over the top of the mountain through Pali Pass. The wind did blow — often. At those times the road was closed to small cars, as they could be blown off the road, and at some places the downward side was quite precipitous.

I was in Honolulu on Statehood Day, which was quite interesting.

* * *

Guam is the southernmost island in the Marianas Island chain. The last one in the chain is Maug (Note: Guam spelled backwards).

During one of my visits to Guam I saw the remains of

Arriving at Hickam AFB, Hawaii, 1950s. L to R: Zimmerman, Jack Ryan, Keith Compton, Bob Smith.

Awaiting takeoff from Hickam. Foreground, L to R: Jack Ryan, Dick Montgomery, Tom Power.

the submarine warning system from World War II. It consisted of a mass of tin cans wired together and installed at certain intervals and depths so that a submarine could not come through an entrance to a harbor without rattling the cans. Lookouts (hearouts?) were continuously stationed at appropriate places.

The liquor distributor for the area gave a large party at his beautiful estate in honor of an admiral who was returning to the ZI for retirement. Weeks later I read that the Customs officials at San Francisco had raided the ship the admiral was on when it docked. The amount of liquor they found and confiscated was fan-

Awaiting takeoff from Guam.

Okinawa, late 1950s. L to R: Zimmerman, XXX, Bob Smith, XXX, Bedwell.

tastic. Since the retirement orders for the admiral had been published, the Navy only inserted a reprimand in his 201 file. But Customs levied a heavy fine on him. The admiral paid quite heavily for his years of luxurious living in the Pacific. Those of us who had been at the party knew where he had obtained the liquor.

Bermuda

On one trip to Omaha from Europe the senior general aboard had decided that after refueling in Bermuda we would spend two nights there. As he got off the aircraft, he was handed a message from General LeMay which said to get to Omaha post-haste. General LeMay had anticipated his boy.

Once, landing at Kindley at about 4 or 5 in the afternoon, we were scheduled to stay across the bay at the lush Castle Harbor Hotel. We had purchased some spirits on the base, very inexpensive. My balcony provided a fine view of a beautiful golf course with the bay beyond. We ordered soda and ice from room service, and I computed that per ounce the soda cost more than the Scotch. Therefore we could not afford to drink weak drinks. A favorite local drink was the swizzle, an authoritative concoction made with rum.

Berlin

During the Berlin Airlift General LeMay decided some additional runway lights were needed for Tempelhof. Some had to be erected in a cemetery. General LeMay checked with the city fathers of Berlin, who said, "Go ahead and get the lights in. They're vital to the undertaking. And if a few graves have to be moved, why a few graves will have to be moved." The moving job was carried out with dignity and reverence. The Russians, snapping pictures, had a field day: "Just take a look! See how the wicked Americans are desecrating the ancient graves. This is what happens in the American Zone."

Some think that propoganda is based on lies. Often it is based on half-truths. However, it can be based on the whole truth: there was a two-horse race, an American horse and a Russian horse. The American horse won. The Soviet news agency said that the Russian horse came in a strong second, and the American horse finished next to last.

From the Russians, General Twining, Chief of Staff, USAF, received and accepted an invitation to visit Russia. USAF officers who accompanied him included General Power, then Deputy Commander, SAC, and Maj. Gen. William H. (Butch) Blanchard, Director of Operations, SAC. (When General LeMay called Butch to his office to tell him he had approved his going, LeMay said, "But no liquor and no women." Butch replied, "That's exactly what my wife said.") General Power and Butch told many stories upon their return. One was about a frightening ride to Stalingrad aboard a USSR aircraft. They were shown the new Stalingrad with pride. Before the war Stalingrad was a dirty, poor industrial city along a river. Now it was new, modern, and attractive. Some of our decision-makers, who have led sheltered lives, do not comprehend the implications of the Soviet achievements in rebuilding their destroyed cities much better than before. What do the Soviets care about the masses of people?

* * *

During the second tour of General Lucius Clay, Sr., as Commandant of Berlin, I was sent on a special mission to Berlin to see him by the USAF Chief of Staff, General Thomas White. Special dispensation had to be arranged. It was at this time that the East Germans (really the Russians) had said they were going to shoot down even commercial planes flying into Berlin. I did not know what was going to happen, but I was at the window with my camera ready. There were only two other passengers on the Pan-Am flight. The winds were such that our approach was over East Berlin. Below, practically no cars, very few persons visible, wash drying on the line the most evident activity. Across a fence (West Germany) much activity, cars going in every direction, many people, a plethora of activity.

* * *

I saw much of the Berlin Wall. Because East Berliners first jumped out of windows and off roofs, the Russians bricked shut the windows and put barbed wire barriers along the roofs. When this proved not to be foolproof, they evacuated every building, area, and facility for 300 yards behind the Wall. There were many flowers on the sidewalks on the Western side commemorating those who had leaped to their deaths. In the suburbs, a big fence separated former close neighbors into East and West. Tree platforms all along the Berlin Wall on the Eastern side, within several feet of the Wall, manned mostly by very young East Germans in uniform. I did not know what would happen but I told my driver to stop. I pointed my already prepared camera at two young armed East Germans on top of the Wall. They ducked behind the Wall, which told me quite a story.

* * *

At the time of the Russian blockade of Berlin, the question was: could the Allies supply Berlin by air? The Soviets set up a study group to explore this question. In February 1948 the results were reported to the Russian member of the Supreme German Council. Leon Uris in *Armageddon* reported that the following was included in the briefing:

Attempts at major air transport of supplies have usually fallen short. Three or four cases were presented.

History shows that one great air-supply which succeeded was by the Americans in the India-China Theater, the Flying of the Hump; but that operation did not parallel the problems of supplying a city the size of Western Berlin and its two million civilians.

The main cargo would have to be coal, and no one knows how to fly coal.

The entire strength of the United States must be put behind the effort. The British are short of crews and aircraft and are in a weak economic position. The French will make speeches about French national honor, but can make no contribution.

To meet the required tonnage, an aircraft carrying ten tons must unload at Tempelhof and be on the way back to the Western Zone every 11 minutes, 24 hours a day. In the winter months the weather should defeat the operation. The weather can be below minimums 60 to 80 percent of the time. You can depend on Berlin to be shut down 50 percent of the time.

The Moscow group concluded that it was impossible for the West to supply Berlin by air.

In June 1948 General LeMay had been Commander, United States Air Force in Europe, for six or seven months. General Clay asked him if he could haul some coal to Berlin. "Sure. We can haul anything. How much coal do you want us to haul?

General LeMay flew a load of coal to Berlin around the first part of July, not the first delivery. He believed he had to make several runs to see how things were going. Some called it LeMay's Coal and Feed Company. In all, the Lift carried 2,223,000 tons of fuel, food, and other supplies into Berlin, five pounds per day for every Berliner. That's 4,446,000,000 pounds of food and other supplies. This is simply one example of many of General LeMay's career of doing the impossible.

The period 1947-1948 in Germany was described as "dicey" by a senior commander in USAFE at the time.

There were great numbers of Army troops, but they were not combat ready. Discharging men was a primary responsibility of the Army.

General LeMay as Commander, USAFE, had two tactical units assigned — the 60th Tactical Carrier Wing, commanded by then Col. Bert Harrison, and the 86th Fighter Wing, commanded by Col. Clarence (Curly) Edwinson.

General LeMay immediately recognized the danger posed by the Russians. On his first Christmas at Hq USAFE, Wiesbaden, the two tactical units were alerted about 0300. They got all planes ready and all crews on board with engines running. But it was not a smooth operation. The crews didn't get home until about 1900. The families eating cold turkey for Christmas dinner, could not understand the exercise. They soon learned, however, and alerts became commonplace, exacting, and efficient.

During that time the dependents were required to be prepared to get out of Germany on their own. They had evacuation kits with things in them for the wives and children for three days. They were to head for Switzerland. The men were encouraged to teach their wives to shoot pistols, and most wives took the training. There were no casualties.

General LeMay set up the Berlin Airlift in June 1948 and decided on the maintenance program, the basing system, and nearly all facets of the entire operation. Colonel Harrison's 60th TCW was the first one on the Airlift, and it performed alone for about three weeks. Then the 61st TCW from Rhein-Main was brought into the operation. Colonel Harrison's men flew the construction equipment to Berlin to build two more airfields, and they evacuated many of the elderly and children. They were harassed by Russian fighters and ack-ack searchlights, but they had an excellent safety record.

General Joe Smith was headquarters commandant at Wiesbaden. General LeMay sent for transports from the United States and put General Smith in charge. Then he asked for General Bill Tunner, who had run the ferries over the Hump. Eventually General Vandenberg, Chief of Staff, USAF, was prevailed upon, and General Tunner arrived five or six weeks after the Lift had started.

Curly Edwinson was a great combat leader and had in the 86th FW some of the finest pilots there were. General LeMay in *Mission with LeMay* describes some of Curly's activities.

Late in 1948 General LeMay left Germany to return to the United States to take command of SAC. General Tunner continued the Airlift to its conclusion.

International Forum of Operations Research Societies (IFORS)

A forum was established among the military Operations Research organizations of the world, and meetings have been held every three years since the first one at Oxford University, England, in 1957. I am one of four or five people in the world who attended the first six meetings.

At Oxford we were billeted in the university dormi-

In England, 1950s. R to L: Wood Rigsby, Keith Compton, Jack Ryan, Butch Blanchard, Tom Power.

Hq USAFE, Ramstein, Germany, 1960. Ross Thackery, Chief, OA, USAFE, at right.

World Wide OA Conference, Puerto Rico, 1959. L to R: XXX, Bill Greene, Percy Carr, Zimmerman, Mart Schwartz.

tories. I was fortunate: I had a 25-watt electric bulb in my room, in Old Bailey. The W.C. (water closet, i.e. toilet) was just across the hall. Some friends in other dormitories on the third floor had to get to their W.C. by going down three flights of stairs, out of their building, and into another building. In the dark, dirty basement of my dormitory was one bathtub for the entire building.

We ate in the University hall. Anyone who has not eaten honest-to-goodness English food, prepared by Englishmen for Englishmen, do not know how lucky they have been. Certain restaurants in London catering to tourists do not count. A SAC base was 10 miles away. Going there occasionally for a steak and shower saved my life.

The second IFORS meeting was in 1960 at Aix-en-

World Wide OA Conference, Langley AFB, 1960. L to R: Bill Blaylock, Chief, OA, ADC; Carroll Bryant, Chief, OA, TZC; Zimmerman; Rod Smith, OA, SAC; Roy Brothers, Retired Chief, OA, USAF.

Senior Commanders, World Wide USAF Commanders Conference, Puerto Rico, 1960.

Provence, France, a lovely town in the French Riviera area. It has many attributes. On a free afternoon I visited Marseilles, including the Nôtre Dame cathedral there. It is on top of the highest hill in the area, overlooking the city and the Mediterranean. On that particular afternoon the wind was blowing with a terrible force. A nun, in her flowing robes, was caught by the wind, and was being blown toward the precipice. I actually tackled her, and we fell to the cobblestones together. Several sightseers, hanging on for dear life, who had seen the episode, applauded. The nun said some very nice things, in French of course. While on the cobblestones I folded her robes tightly around her, and as we got up, I held them firmly in place. She gave me a big hug and a kiss on the cheek. I shall never forget the look in her eyes.

After the meetings I spent half a day at St. Tropez, the resort for the European elite, including those in the entertainment industry. Even then topless lovelies adorned the beach. I had lunch with Jack Paar and Charlie Weaver. We were at a front table at a sidewalk cafe next to the beach, so the view of the scenery was quite good. Charlie Weaver without his make-up and bald Jack Paar without his wig looked quite different from their TV personalities.

I could tell much about Oslo, Norway, 1963, the land

World Wide USAF Commanders Conference, Puerto Rico, 1960.

World Wide USAF Commanders Conference, Puerto Rico, 1962.

of the midnight sun and free love, but I shan't.

The meeting at MIT, 1966, was not the most outstanding one. One of the amusing aspects, to me, was that I never saw the three Russian delegates at any of the sessions. They were much in evidence in the lobbies, halls, and courts talking with anyone who would talk with them, and they must have taken hundreds of pictures. One of them would go and stand in front of a building a block away: to get the entire building the photographer had to be that far away, and taking "tourist picures" was a pretense. At some distances they would not be able to tell if the guy in front of the building was Russian or Chinese. But they got pictures of the buildings.

My wife accompanied me to Venice, Italy, in 1969. My first meal there consisted of French-fried fish eyes and sea horses. After Venice we visited Rome, Naples, Athens, Vienna, Zurich, and London.

Conference I.D. card.

The sixth IFORS was held at Trinity College in Dublin, Ireland, in 1972. I had been in Ireland three times but never to Dublin. After a week Dublin became dear to our hearts. It is hard to understand why; it certainly was not the lousy weather. It must have been the people.

International Forum of Operations Research Societies, Venice, 1969. Front row: Zimmerman, Ed Berdahl.

Chapter 11

Flight Experiences

With over two million miles in the air, it is not surprising that I have had some interesting flight experiences. I've had 40 flights to Europe, 7 to the Far East, to the Middle East, several to Africa, South America, various Pacific Islands, Iceland, Greenland, Alaska, and many within the Zone of the Interior. The events are not in chronological order.

* * *

My first flight was at Walla Walla, Washington, in a four passenger aircraft of some kind, piloted by the manager of the Walla Walla Airport. He had engaged me to teach ground school (meteorology, structures, maintenance, and so on) in his Civil Air Patrol pilot training program.

While stationed at Hq IX Tactical Air Command at Middle Wallop, England, I looked out of my office window and saw a person in flight clothing. I asked my sergeant, "Isn't that a girl?" Then she turned around and I recognized her. Bobbie Sandoz had been in my CAP course at Walla Walla, and now was a ferry pilot for the RAF. She was flying a Spitfire when she had some kind of trouble and landed at the first base she saw. We had lunch together at the officers' mess.

At another base in England I ran into a lad who had taken the same course from me at Walla Walla. He was a local Walla Walla lad and both my wife and I knew him, his brother, and his mother. His father had passed away. He was shot down on his first combat mission, but he survived as a POW to return home. His brother had been in the same program at Walla Walla. He was shot down somewhere in the Pacific and presumed lost, but after the war he was discovered as the "king" of the natives on a small uncharted island. Imagine the emotion of their mother: after such initial bad news, both of her sons returned home.

* * *

My first flight to Europe was in late May 1943 in a clipper flying boat — unpressurized, of course, and able to land only on water. (This did not seem a bad idea on my first flight across the Atlantic.) From Botwood, Newfoundland, to Limerick, Ireland, 11 hours. After we passed the point of no return we got into a terrible storm, and after we landed the ground would not hold still for me for several hours. I heard the copilot say it was the worst weather he had ever experienced.

* * *

The C-47 (DC-3) was labeled a Gooney Bird. I once flew from Washington, D.C., to Anchorage, Alaska, in a C-47. Some of you will comprehend what an experience and ordeal that was.

Once I flew as the only passenger on a United Airlines DC-3 from Denver to Washington, D.C. Instead of announcing that Flight X was leaving, they announced the Zimmerman Special was leaving. Before landing at Chicago (no passengers got aboard there) and Washington the stewardess folded up the blanket and put it and the pillows away.

Anyone who has seen a bird Gooney Bird understands why they are so named. They land at a speed much faster than they can run, so on each landing they go ass over applecart. On Wake Island I watched one taking off. He paddled down the runway and got airborne, but not in time to miss the hanger at the end of the runway in the direction and distance he decided to take off. He smacked into the hanger near its top and plopped to the ground. He went back and tried it twice more, with the same result.

* * *

During World War II I flew from Middle Wallop to Land's End on a small British aircraft with a British RAF pilot. It was excitingly scary. I differentiate between excitingly scary and frighteningly scary. In blacked-out England there were no lights, not even on the runway at Middle Wallop. Upon landing I made some comment about the very low flying, to which the British pilot said "I could not fly 'on the deck' because of the rivets in the bottom of the aircraft."

On a flight from London to Prestwick, Scotland, the weather was lousy. The pilot flew "with one foot on the ground." The only passenger who got airsick was a B-24 Liberator pilot. He had completed his tour of duty and was on the way home. It's quite different for a pilot when he is not driving the aircraft. This should induce some pilots to have some empathy for mere passengers.

* * *

During World War II the number of combat missions constituting a tour of duty for the crew of a given aircraft type was determined by computing the number of combat missions, according to the attrition rate, that would give the crew a 50 percent chance of survival. Yet some people today are horrified at a potential 50 percent attrition rate. Though it wouldn't be that much, regardless of what militarily inexperienced armchair strategists say, if it were 50 percent the crew would most probably make only one combat mission. On that one mission the crew would drop many more equivalent tons of explosives than that totally dropped during the 30 or 50 missions required in World War II.

* * *

About five minutes from takeoff from Anchorage, Alaska, the KC-97 aircraft started vibrating in a violent, frightening way. General Bill Fisher, who was sitting with me, ran up front. The vibration stopped immediately. Upon return he said the crew had forgotten to close the cowl flaps. Later a commercial version of the same aircraft crashed about five minutes after takeoff from Seattle for the same reason. All aboard were lost.

At Kwajalein in a driving rain I had two refused takeoffs in a C-118 (DC-6). This was not the most comfortable feeling one could experience.

On takeoff from the Azores in a C-54 we barely missed the top of the high hill at the end of the runway. The pilot circled and landed immediately. Besides the fact that we were heavily loaded, the ground crew had forgotton to take the pin out of the nose wheel. Hence it could not be retracted. The added drag caused problems in this crucial phase of the flight.

* * *

When living in Omaha I planned to go there one time on SAC's AJAX Courier, then a C-54, from Wright-Pat AFB. We were supposed to take off at something like 8 in the evening. When I got to the flight line I was told there would be an hour's delay. I went into the crew flight-planning room, and found that in Omaha the dewpoint was within 2° of the temperature, and the alternate was Colorado Springs. I went to the desk and asked them to take my name off the manifest, that my plans had changed. I went to the VOQ and had a good night's sleep. Upon awakening the next morning I phoned to ask if the AJAX was still there, and if so, its scheduled departure time. The answers were yes, and in one and one-half hours. I asked them to put my name on the manifest. When I got to the flight line I learned that the AJAX had taken off the evening before, flown two-thirds of the way to Omaha in terrible weather, and then had turned around and come back. What a sorry-looking bunch of passengers!

* * *

A KC-97 took 10 to 11 hours from Paris to the Azores. Once when arriving at the Azores we had to circle an hour before landing, because of weather. On the same flight another 10 hours to Bermuda, and again we had to circle for an hour because of weather. In each instance there were not too many alternates: the number of alternates was zero. One cannot describe that as particularly pleasurable.

* * *

When taking off from Kindley AFB, Bermuda, in a KC-135, it was discovered that the ground crew had failed to put the cap on the wing fuel tank after refueling. Fuel was splattering everywhere, which was not the most comfortable situation to be in. We went around and landed without incident. Every meat wagon (ambulance) and piece of fire-fighting equipment available were lined up for our landing, and sped along with us to our stopping.

* * *

Pre-jet I would often go to Europe on Military Air Transport Service. Upon arrival at Frankfurt, Germany, the Commander, USAFE, would provide a C-54 with crew, including a galley and a WAF, for my use. On one trip I went to Greece, Turkey, and Libya. At Incerlik, Turkey, something looked different to me. Then I realized what it was. All the fighter aircraft were facing away from the flightline and hangers. Some would have to taxi on grass a bit to get into position for takeoff. I asked the senior USAF colonel, who was squiring me around, "Why?" He replied that those aircraft had live ammunition in them and were flown by Turks.

On the way back trouble developed in one of the four engines. The flight plan included a stop at Aviano, Italy, in the north at the foot of the Alps. No trouble making it there on three engines. On the ground the faulty engine worked perfectly. They did some checking and adjustment. My aircraft commander said I had two choices.

We could stay there until a new engine was flown down from Frankfurt, or we could climb to altitude over the flat part of Italy, make a trial run at that altitude, and if things seemed okay we would go on to Frankfurt. The Alps we had to fly over were over 15,000 feet high. I chose the second alternative, and we got to Frankfurt without incident.

* * *

In the early days of Rand they had a policy that no more than three Rand employees could fly on the same aircraft at the same time. I don't know what their policy is today. I thought of this when returning in a KC-97 from Eniwetok to the ZI. Aboard were General LeMay, Maj. Gen. Cam Sweeney, Director of Plans, SAC some other members of the SAC staff, and I. Plus Edward Teller, Herb York, then of Livermore Labs, Norris Bradbury, Director of Los Alamos Labs, and others.

* * *

When I was flying over the Greenland Ice Cap in a KC-97, General LeMay called me from Omaha to tell me that Rod Smith's and Stan Lawwill's supergrades had been approved. Hitting the squawk box to my office was one thing, but this was another.

The first time I called home from an aircraft was when flying over Germany in a KC-97. It was about eight o'clock Saturday morning, Omaha time. I called via Military Affiliate Radio Service (MARS), which patched into the commercial phone system. My daughter answered. She was probably about 8. She yelled, "Mommie, Mommie! MARS is calling."

* * *

I have a picture of the Equator taken from the air over Brazil.

* * *

In the early days of the operation of the Boeing 707 I made an uncontrolled landing at San Antonio in a commercial 707. On approach the pilot permitted it to stall. It bounced first on the wheels on one side, and then the wheels on the other side. After a few more it stopped off the runway. Fortunately the baked clay ground was about as hard as the runway. The landing bent the fuselage to the extent that the back door would not open. The colonel who met me said it was quite a sight from the ground.

* * *

SAC had just received several U-3 four-passenger aircraft. I was being taken from Omaha to Minneapolis. A while after takeoff I tapped the pilot on the shoulder, held up a cigar, and said, "May I?"

He said, "Yes, but don't set the aircraft on fire, as we have no parachutes."

I replied, "Major, even if we had parachutes I would not set the aircraft on fire."

* * *

In flying commercial from Denver to Omaha in very bad weather the passengers all had their seat belts fastened. The stewardesses were doing the things they usually do. We hit an air pocket. Suddenly the stewardess near me was plastered against the ceiling. On the way down she probably made a mistake in trying to lessen her fall with her arms. Both of her wrists were broken.

* * *

Long before jets I was flying from Colorado Springs to Albuquerque. The night before, a general and I had been out on the town; even sitting at my desk the next day I would not have felt too chipper. The weather was lousy, and we had a stop at Sante Fe. That was the closest I have come to getting airsick. The thing that saved me was the purple sweater sitting next to me. It was her first flight, and she was terribly frightened. Reassuring her saved me.

* * *

On a commercial flight I asked a stewardess if she had a match. She looked at me for a moment and said, "I don't think so, but I can give you a light."

* * *

Modie and I made an inadvertent mistake. We landed at the new Athens International Airport the first day it was operational. What a mess! Landing at a Greek airport that has been operational for a long time is problem enough.

* * *

Pre-jet to get from Omaha to the West Coast one always stopped at Denver and Salt Lake City. The airport at Salt Lake City is at the foot of the high Rocky Mountains to the east, so a considerable descent is required coming from the east. On one flight in a DC-6 the pilot nosed the plane down so sharply that we would not have stayed in our seats if our seatbelts had not been fastened. Unfortunately, he did not flare out soon enough and the nose wheel hit the runway with a tremendous force. Fortunately the wheel held and we came to a safe stop, but it was damaged sufficiently that they changed aircraft for the continuation of our flight.

* * *

My first flight in a Boeing 747 was to Europe and back. On the way over I figured it was 28 years to the day when I had made my first flight across the Atlantic in a Boeing Flying Clipper. On the way back my military assistant, without my knowledge, informed the crew that it was my birthday. Tradition provided a free bottle of champagne. With something like 10 stewardesses aboard, it was a pleasurable trip back.

* * *

There were only two flights in my two-plus million miles when I was genuinely frightened. Both were in commercial aircraft. One was across Lake Michigan in a terrible thunderstorm. The other was across Indiana and Illinois to Chicago in a British Viscount through a thunderstorm.

Chapter 12

Strategic Air Command Senior Personnel

Brig. Gen. Robert O. Cork, Hq SAC, 1953.

General Thomas S. Power

Brig. Gen. Thomas S. Power became Deputy Commander, SAC, on 26 October 1948. (He was promoted to major general in 1949.) He asked me to accompany him on his first trip to Europe as Deputy Commander. I had just returned from Europe, so I sent my deputy, Stan Lawwill, in my place.

After that I went on every overseas trip and many in the ZI with General Power, both as Deputy Commander and CINCSAC. Overseas he always took the same four or five members of the staff. We conducted "think" sessions, discussing issues and problems that we probably would not have taken the time to discuss at SAC Hq. We visited overseas SAC installations, and at night we would have no-holds-barred discussions. General Power was only another participant. He insisted that we not be influenced, or remain quiet, because of his position.

There was some time for recreation, particularly golf. I have never played golf. I used to say, when asked, that what spare time I had I would rather spend with my family. One morning I awoke and realized that the other three members of my family played golf. I never did. When General Power was playing golf I would watch them tee-off at the first hole. Then I would usually explore the countryside and meet them at the nineteenth hole when they were finished.

Maj. Gen. Butch Griswold, Hq SAC, 1950s.

Col. Richard T. King, Hq SAC, 1950.

Col. Alfred F. Kalberer, Hq SAC, 1950.

Maj. Gen. August Kissner, Hq SAC, 1953.

One day at St. Andrews in Scotland there was a delay in their teeing off at the first hole. General Power was reminiscing. He said, "If I had my life to live over, I can't think of much that I would change." Col. Bob Smith said, "If I had four stars I would feel exactly the same way." (Bob Smith retired as a lieuenant general — three stars.)

Particularly pre-jet it took many hours to grind across the oceans. We would have "think" sessions and play bridge on the aircraft. Modie bought me a thick book on bridge by Goren, so we would discuss our hands and look up specific situations in Goren. General Power and I learned Goren together. We both became quite accomplished.

Once on an aircraft with General Power as my partner I bid five diamonds over our opponent's four-spade bid. One of the opponents was then Major General Jack Ryan. I was doubled. To show his disgust General Power laid his head back, pulled his hat down over his eyes, and would not watch the playing of the hand. I made the bid.

General Power had a reputation for being tough on his staff. Any wrath he imposed on them was usually well merited. He had a warm compassion for his troops. General LeMay tells the story in *Mission with LeMay* about the maintenance officer in the Pacific who pounded on General Power's desk. He was filthy dirty, smeared with grease and grime. An aircraft on which they had worked three days and three nights was not scheduled for the mission. He said, "Maybe we were late, but we had troubles. The rest of them haven't taken off yet. By God, she's got to go, General."

General Power replied, "O.K. She is going to go." She went.

When General Power was deputy commander, I was in his office when he received a call from Topeka AFB, where a training program for SAC noncommissioned officers was being conducted. Since it was only for six months, expenses for moving their families were not authorized. A sergeant had decided to take his family at his own expense from California to Topeka for six months. When he arrived there, he was told the schedule had been changed. He was not due now until six months later. The senior SAC man at Topeka, a colonel, had the courage to phone General Power, who phoned the commander of Air Training Command. The words flowed in an elegant and convincing manner. The sergeant was included in the current class without delay.

I remember an effective sign, for me, on the bulletin board of the officers' club at Topeka AFB. It said, "At 70 mph it takes nine minutes to drive to Topeka. At 45 mph it takes fourteen minutes. Why don't you leave five minutes early?"

Another time when I was in General Power's office, he started to light a cigarette. He said, "This is a nasty habit. I shall not smoke again." He did not. I was careful never to smoke cigars in his presence.

Maj. Gen. J. B. Montgomery, Hq SAC, 1953.

General J. P. McConnell (USAF Ret)

This former Chief of Staff, USAF, was one of the

Maj. Gen. J. P. McConnell, Hq SAC, 1950s.

Maj. Gen. Archie Olds, Hq SAC, 1950s.

most competent, brilliant, dedicated, and considerate gentleman I have ever known. We became quite close over many years. He would give the impression of

Maj. Gen. Robert Terrill, Hq SAC, 1950s.

Maj. Gen. Thomas S. Power, Hq SAC, 1950s.

Maj. Gen. Jack Ryan, Hq SAC, 1950s.

Maj. Gen. Walter Sweeney, Hq SAC, 1953.

Maj. Gen. David Wade, Hq SAC, 1950s.

Col. Kingston E. Tibbetts, Hq SAC, 1950.

being an Arkansas country boy, which he had been. But if discussants did not know what they were talking about, were not completely prepared in their subject, or tried to pull a "snow job," they got full benefit of his sharp, considered judgment, in a manner they were not likely to forget soon.

I first became associated with General Mac when, as a major general, he became Director of Plans at SAC Hq in Omaha. I earned my spurs with him after the following incident: he was explaining a recommended position to General LeMay, and at the end he asked, "Don't you agree, Zim?" I said, "No," and told why. You know how most people would have reacted to that situation. But with General Mac it made it for me. He went beyond the call of duty to help me. He gave me opportunities. He sought my advice and counsel.

A similar occurrence resulted in my being close to General Mac's predecessor. Col. Cam Sweeney became SAC Director of Plans soon after General LeMay took over as commander. One day he had four or five of his senior people, and me, in his office. He described a course of action he wanted to take, along with a suggested approach. Then he asked if anyone disagreed. None of his staff did. I have never been very talented at concealing my true feelings. Apparently Colonel Sweeney noticed this, and asked, "Do you agree, Zim?"

Col. Hewitt T. Wheless, Hq SAC, 1950.

"No."

"Why?"

"Because it is unethical, if not dishonest."

How would most military men react to a civilian

saying something like that in front of his senior staff? But Cam Sweeney was different from most men. We became close. He progressed in his career to four stars, but then had a most untimely death.

General LeMay had the knack of getting such people around him. At this point I should insert that in some quarters General LeMay had the reputation of having only "yes" men on his staff, and that he personally was quite arbitrary. Those who were fortunate to work closely with him know how untrue, absurd, and unfair such allegations were. General LeMay's staff could support his positions with vigor, because we had such a large part in formulating them. Arbitrary? Most of the time he had ready answers for those who brought him the best idea since sex, because he had thought of it long before they had, and had carefully thought it through.

General McConnell told me one day that he was going to have a hernia operation. He said that it wasn't because he didn't trust the doc, but none the less he was going to paint the family jewels purple. He came back to work too soon after the operation, so had additional difficulties. One day he said, "Zim, if the doc ever tells you that you have a hernia, you tell him that's just fine — that's the way you like it."

At Hq SAC General LeMay carried out the custom of having the senior staff in his quarters on New Year's Day. For one of these events General Mac donated several quart fruit jars of moonshine from the hills of Arkansas. General LeMay required everyone, including the wives, to take a swig. He explained that he wanted to see who had indentations on their face a quart fruit jar's distance about the mouth, so he could tell who often tipped quart fruit jars.

One evening while at Homestead AFB, south of Miami, we went to Marathon Key to look at some property General Power owned. General Power and I sat in back. General Mac sat in front with the driver. As we started, General Power said, "Mac, throw that stinking cigar away." General Mac put it out and disposed of it. Further down the road General Power said, "That must have been the world's worst cigar. The damn smell is still making me almost ill." When we arrived at Marathon, as we were getting out of the car, General Power discovered that somehow General Mac had failed to get the cigar out of the car. Instead, after he put it out, the stub somehow resided at General Power's feet.

We stayed at Homestead that night. About 2 a.m. General Mac was notified that one of his C-135s had crashed, with all aboard killed. The procedure was to notify CINCSAC immediately when an aircraft was lost. General Mac decided not to bother General Power at that hour. But Omaha did bother General Power with the information at that hour. Sparks? Yes sir!

With four stars General Mac was Commander European command then in Paris. I visited him there and

Col. Stuart P. Wright, Hq SAC, 1950.

congratulated him upon becoming the next Chief of Staff, USAF. There had been no announcement and I really didn't know. He replied, "Zim, you know I'm a country boy. The big cities are too much for me. I want to be CINCSAC."

I said, "Let's shake on this agreement. After you are named Chief of Staff, I'll give you a week to get your feet on the ground, and then you owe me a night on the town.

It was in 1974 that General Mac told me he was slated to become CINCSAC. However, the recommendation for Air Force Chief of Staff was turned down by McNamara, so Mac's assignment was changed from CINCSAC to Chief of Staff.

After he became Chief of Staff, I would visit him on each trip to Washington. I was with industry in California during those years. On one visit his appointments secretary said that General Mac was at lunch and had a joint Chiefs of Staff meeting at 1:30 p.m., but perhaps I could see him a bit before the JCS meeting. At 1:25 I went to the back door of his office (I knew he went out that door to the JCS), and very soon he appeared. I said, "I simply wanted to say hello."

"What's your hurry?"

"You have a JCS meeting."

"Come on in. I'll tell them I had to talk to Zim."

We went into his office, where he took off his blouse and hung it over a chair. As he offered me a cigar, he said, "You know these are too good for you." "Yes, Sir." At SAC he was always bumming cigars from me, and then would complain bitterly about their quality. On my only trip to Manila, I brought back to him two boxes of his favorite cigars.

I tried to leave several times. He would not permit it.

Finally at 1:50, already 20 minutes late, he said, "I must go." He put on his blouse, we shook, and he left.

In 1973 he told me this story: As majors he and Monk Hunter shared quarters at Barksdale AFB. Mac loved to hunt doves. One Friday evening he told Monk that he'd certainly like to go dove hunting the next morning, but he had no gun. Monk had an $800 shotgun from Italy, which should tell you it was quite a gun. Monk said Mac should use it. Mac demurred, but upon Monk's insistence, accepted.

Just after daybreak Mac spotted hundreds of doves sitting on transmission lines going parallel to the road, but through a field. He was having great fun blasting away, when an unmarked car stopped. A man got out and said, "Young man, what are you doing?"

"Shooting doves."

"Do you not know that it is unlawful to shoot doves on public property?"

"I know those transmission lines are on private property."

"Who do you think owns the land on which you're standing? I must take you to the Justice of the Peace."

"If that's necessary you go along and I'll join you there."

"Young man, it will be no trouble at all following you."

Mac had only a couple of dollars. He asked, "Is the Justice tough?"

"It depends on how he's feeling. However, he has not failed to fine a single individual I have taken to him."

After giving Mac quite a lecture, the Justice fined him $50. He said, "I'll have the doves." With sadness Mac handed him the sack of doves.

"I'll have your license." He tore it into shreds.

"I'll have your gun." Mac's chin fell to the floor. Monk's $800 gun!

Permitting Mac to stew in his own juice for a while, the Justice said, "Come back at the end of the dove season, and you may have your gun."

General Mac — a truly great professional and gentleman!

Chapter 13

Strategic Bombers and Security

B-52

In 1955 the 93d Bomb Wing at Castle AFB was the first wing to be equipped with the B-52 Stratofortress. This was quite an accomplishment, because during its development the B-52 was nearly canceled four times. Each time, a visit to Washington by General LeMay saved the program.

Since B-52s must remain operational into the 1990s, where would we be without them? Not only because of their past accomplishments, but because of their potential for the future. But it is sad that the United States Air Force and the nation must depend on the B-52 for several more years. The last B-52 was delivered in 1961, so the *newest* one will be 29 years old in 1990.

There were attemps to get replacements. The B, RS-70 and B-1 stories are well known. McNamara canceled the B, RS-70, and President Carter the B-1. Fortunately, President Reagan has included the B-1 and a new strategic bomber in his military budget.

Armchair strategists, incompetent decision-makers, politically motivated decision-makers, and naive and inexperienced Whiz Kids have had a vendetta for decades against manned bombers. Their arguments about vulnerability simply cannot be substantiated. But they prevailed until President Reagan.

I once gave a talk before a large group of high level Army, Navy, and Air Force personnel, which included the Assistant Secretary of Defense for Systems Analysis (the Whiz Kids). I said there are two instances where the objectivity of an individual cannot be questioned. One is a self-appraisal. The other is if one holds a high enough position in the Department of Defense.

My knowledge of bomber penetration includes more than analytical studies. My son, Maj. Carroll L. Zimmerman, Jr., is in charge of B-52 contingency training for SAC at Castle AFB, California. On two tours to Southeast Asia as B-52 Aircraft Commander he had 116 combat missions. He was awarded the Distinguished Flying Cross. During Linebacker II no B-52 aircraft commander had more missions over North Vietnam than he.

Comments from the 1960s

Since I'm writing my memoirs, I'm listing the date when I wrote the following papers, and I am including them here exactly as I wrote them. Four four-star generals agree that much in them is relevant today. This is sad, as it shows we are not only not staying ahead, we are falling behind:

Security in the 1970's	(1964)
AMSA, B-1 and Deterrence	(1966)
AMSA, B-1 Penetration	(1966)
Command and Control	(1960)

Security in the 1970s: What Kind and Whose? (1964)

[This paper was written in 1964 and given to the Ninth Annual Aerospace Symposium of the Air Force Association Utah Wing at Salt Lake City in February 1964. I was a member of a three-man panel discussion. The others were Maj. Gen. Jerry Page, then in Plans, Hq USAF, and a Rand representative. I was Director, Aerospace Analysis, Missile and Space Systems Division of Douglas Aircraft at the time.]

Your symposium chairman, gentlemen, is quite optimistic. He has asked the three of us to paint a picture of a world not yet seen. He said, "Each of you is asked to give a 15-20-minute prepared presentation on conditions, problems, actions, etc., affecting national security in the 1970 time period as you see it from your vantage point and profession." This for a full decade that, even at this moment, is more than half a decade away.

I am grateful that we were encouraged to rely on the tools of our trade (our "profession" he called it). I believe that the perspectives of analytic method enable one to see, however uncertainly, across so awesome a time span. So I shall approach the requirement of prophecy with hope. Toward the end, I shall impose upon your patience by sketching some roles of my own industry in the enfolding environment of effort which lies before us.

"Security" we may first observe is a plain word. A man uses it in describing his job, or his income, or his insurance. In our context it is a *high* word. And it evokes the most urgent kind of public and patriotic responsibility. Despite its plainness and familiarity, security is a *vague* word. Would it describe the condition of the Nation, a condition where peril was either absent or mastered? Would it indicate a stable and controlled set of circumstances where we had nothing, or almost nothing, to fear?

Or — should the term be employed in a large and extended sense to signify something closer to "freedom" for a way of life, or the pursuit of expertise, for a ready and human commerce with our neighbors? Is security a fixed, stable possession? Which can never be eroded by shifting climates and environments? By fluid events and ideas which determine the proximate medium in which security is held and enjoyed?

The term, quite obviously, is abstract. But genuine security never occurs in the form of its abstractness. Security, in a very literal sense, has to be secured and kept secured. It is an operational word and is rather a description of what one does. It is shorthand for management of events and affairs. It finds real meaning only when viewed as a prescription for management and for the control one must exercise in view of goals and objectives.

Because it deals with goals and because it deals with management, it is a "foresight" word. It is an effort word, an action word, and it implies decision, design, and determination. In effect, then, when we contemplate the problems of "Security in the 1970s" we must examine very carefully the precise kind of issue we are bent on examining. We are not pre-writing history quite so much as transcribing the formula for our own endeavor.

So the question will be less about "what world awaits us?" than about "what kind of world do we wish to build or help build?" Our investigation will require us to advert to the fact that what we do now constructs the world of the 1970s. And it surely will command us to remember that both the world of our future and the security we achieve at any time interval are, both of them, functions of intelligence, judgment, sincerity, and will.

It would, of course, be quite possible to "take a squint" into the 1970s. We could simply assume that hardware, the hardware both on the drawing boards and in being, could somehow be a clue to at least one component of that future world. We might simplify our prediction problem and take for granted that the Soviets will remain a serious, or our only serious, military threat. It would make some sort of realistic sense to say, further, that Red China may, sometime rather soon, be a nuclearized and predatory analogue for Stalin's Russia. [Remember this was written in 1964.]

If one persists in squinting, he might round out this crude picture by noting that the rising nations will continue to rise. He could add that trouble in the Caribbean and Latin America will get worse before it gets better. These items once selected could be extruded forward and, to that extent, some scenario of the 1970s might be said to be in hand. Such a picture in whatever setting would truly be a squint. It might be less than honest. It could be dangerous.

Clearly, if we are to take the long look, we must consider items like those just listed. We must, in fact, bracket and question almost every element of this contemporary time which in being so close to us seems so real and so enduring. We must go even further and lay out every probable and conjectural element which the best foresight may say will be present in the 1970s and which will have some security relevance for us. Yet, I submit that our security problems are operational in nature, and much more than this. We have to list all of these items, but we must look at them with a managerial eye.

The USSR and Communism may be matters of concern to the publicist and the historian. But our concern with both turns on their inner dynamics, on the world they are creating, or will create. And our insistent question has to be: HOW DO WE SECURE OUR SECURITY?

Red China must be viewed on any reading as having an enigmatic and ominous meaning for the next 20 years, or so, of human history. When we hear the echoes of propaganda, influence, and subversion in Africa, the Caribbean, and Latin America, as well as SEAsia, we must recognize management implications which stand quite apart from even a fact of major importance such as the generation by Red China of a genuine nuclear capability, with or without the help of her neighbor.

Without any question we must analyze with penetration and care the ongoing dynamics of the rising nations. Perhaps, too, we must refrain from the facile device of pinning Communist labels on all manifestations of change. Surely we must accord to the native dynamics of these movements the same diagnostic care which the Communists do. It's probably true, also, that the Near and Middle East should join Latin America, SEAsia, the South Pacific, and Africa as being objects of our more intensive concern.

All the items so far mentioned are the staples of every generalized discussion of our national security. If we add a few more routine items, we have all of them: the United Nations, regional groupings, the shifting power balance, well-muscled anti-Communism, and technical but sophisticated discussions of strategy, deterrence, and force structures. For the most part, I submit, perspective is lacking in these discussions. I believe that our most insistent responsibility is TO LAY OUT A MAP OF CONTROL-POSSIBILITIES, and to scrutinize each item on it. In this perusal, our most imperious requirement is that we maintain the perspective of initiative and management.

Each one of us, I am certain, has his own map of mandatory elements — those items which in his eyes both determine the future and are, accordingly, the raw material of policy and decision. We might agree that the first such entry would be this: What future form and substance shall we give to the distinctive American experiment — our revolutionary enterprise of human perfection, dignity, and freedom? What is the potential, and intended, shape of the American presence in the world of the 1970s? But, for the moment, let us leave this consideration until the end of our map.

There are, I suggest, three broad bands of mandatory action which, if traversed, will provide the arena for the American dream — an arena of hope, determination, and accomplishment. First we must make our way to the 1970s. A rather obvious point. Yet present concern, especially in the climate of danger, has a way of shortening the range of our perception.

The issue of fear-telescoping prompts me to make four observations:

We are over-concerned with the instant threat and danger, magnifying Soviet power and enterprise to the detriment of the dynamic progress of events.

We are, thereby, over-reactive. We react rather than act. Partially, of course, our accepted "second strike" policy sets us quite firmly in some respects in a reactive posture. Yet, recognizing the proper issue of non-aggressive policy in one quarter need not spell total passivity.

Economy has become some sort of obsession with us. While it is only too obviously true that the economic base of current security postures are, in their way, important values, it is not an independent constant. Efficiency and frugality are to be commended. But in our anxiety to be frugal, let us not substitute for past economic inefficiency something like an irresponsible and long-range insufficiency. The truth is that instant military demand is in fact only a way to the future. But it is the future which is important, and we better get there.

Current strategic concepts seem shaped to a reactive posture rather than, as they should, to the demands of futuristic and mandatory dimension. Arms control quite obviously is a legitimate objective. Likewise the formulae of counter-force doctrine. Plainly, too, we ought not be locked in the bonds of a single crude attitude of response. We need options and flexibility.

We also need the elaboration of a sort of "Policy of anticipation" in which we foresee and pre-think whole ranges of hostile feasibility. We now lack the dimension of anticipatory action.

Our security must give full value not only to the possibilities inherent in our run-away technology, but also to human ambition, enterprise, and dynamics which both forms the technology and follows in its wake. It is, without doubt, one of the curiosities of our time that we should both fear and seek to avoid the implications of technological advance. We fear those things we do not understand. Therefore we must understand the implications and ramifications of technological advance. It will not go away. We cannot, and certainly should not, legislate against invention. On the one hand, we spell out our fears of Soviet "breakthrough in technology" all the while debating how much we should spend to put a man on the moon. Both are more than possibilities to be contemplated. They are imperatives for action. It seems that we, this responsible generation, must face up to the responsibility of anticipatory action in both areas.

Dollars are hardly a reason for risking USSR pre-emption of weapon systems. It appears to be pure ostrich-ism to argue the genuine scientific and technological exploration of space solely on the grounds of narrow economic concern. Any reading of history and human curiosity will assure us that the full exploration of space is no longer a matter of choice. Space exploration hinges upon our scientific knowledge. And the latter is a function of space technology. Space needs its "microscope." Without the latter, the former is impossible. Utilization of space (the peaceful vs the military) is a consequence of technology, resultant knowledge, and national purpose. It is not an ab-reactive obligation.

Furthermore, there are other arenas of national action which rely directly on ideas, science, and technology. Add to these a third area of social change and social dynamics. Here we ought to think of three problems: the "rising" nations, the old nations including Latin America and our European allies, plus our own. Our real responsibility is to lay a diagnostic hand on nativist movements by underdeveloped nations and ascertain their true force and direction. The Communists do. Else — how pre-empt them? A somewhat similar touch is required even with our Allies. What about the Frances and the Englands? France has, we may note, released Red China from diplomatic bondage. England sells buses. What happens when the receiving line grows longer and more crowded? Or the handouts more frequent?

Finally, let us return to the first item on our problem-map: the issue of our own future presence in the world of the future. The entire concern of national security is to guarantee the possibility of a many-faceted American future; particularly of a future which will both tolerate and support our national development as a nation. We must surely give thought to the legitimate extension in time and over time of the fundamental concerns of the American experiment: Courage, enterprise, live and let live, freedom, knowledge, practical concern.

Our resultant scenario for the 70s then, is something more humble and, I hope, more responsible than the word "scenario" might imply. How much are we willing to secure? I have no picture of the 70s. I do have a spectrum of imperatives for the "here and now." Several times I have made the point that what we do now determines the 70s. Thus, in the 70s we ought to guarantee the accomplishment of the following critical tasks:

The elaboration, with precision, of the character of our presence in the future world. This should be non-aggressive. But neither should it be merely passive and compliant. We must choose and help develop the nations we wish to live with.

We should identify and define many possible future world situations and make clear to the world our intended actions, *before they occur*. Then prove to the world the credibility of our words.

A posture of mere reaction to hostile forces and actions seems incompatible with so reasonable but so ambitious a goal. Hence, our first imperative is to alter adroitly our curious current style.

Inasmuch as we are not aggressive by national conviction and policy, let us hold fast to our fundamental axiom of "second strike." But only where defined may our policy be reactive.

Our basic strategic outlook should be characterized by a concern for the prevention of central, nuclear war. Plus the steps that may lead to that — e.g., a conventional war in Central Europe.

We must out-strategize our potential foe. Three areas await exploitation: technology, space, and counter-insurgency.

Finally, we require credible capabilities in all three areas. Where economy is our concern, the only valid issue can be one of timing or staging — not of objective or relevancy. The entire layer of indigenous social ferment requires fresh and resourceful, humble research.

The above items and issues are intimate concerns for the entire aerospace industry. Security is the environment and the milieu of their existence and presence, just as it is for the plainest citizen. Much has been pointedly said about the alleged dangers of the military-industrial complex. This audience, I am confident, needs no comment from me on either the point or the irrelevance of any speculation on this will-of-the-wisp. Whatever the industry can or is willing to do, in the longer perspectives of the 1970s, is just one input — major perhaps, but just one on the larger picture. The concerns of all are for the later and instant tasks, which are both the keys to and the responsibility for that future.

AMSA (B-1) and Deterrence (1966)

The word "deterrence," used in a military sense, is well known to the American public. There are several definitions, varying in degree, but they all mean one thing: to have a military force so strong that it will prevent a would-be aggressor from beginning a war because he knows this force will be capable of dealing with him in a manner and to a degree which he is unwilling to accept.

What kind of war should we be capable of deterring? U.S. military policy is to be sufficiently strong throughout the entire spectrum of possible conflict to deter any level of aggression — from guerrilla warfare or insurgency to and including all-out nuclear war. However, should deterrence fail, the policy is to be strong enough to win on terms acceptable to the United States.

We should note carefully that having a general war deterrent capability was and is mandatory to be able to generate and apply the strength to win in the very difficult current Vietnam situation. Having a credible general war deterrent capability in the future is mandatory if we are to be able to meet and handle successfully any possible future threat less than general war. A credible general war deterrent capability makes unlikely the escalation of lesser conflicts to all-out nuclear war.

Thus this paper is largely concerned with the deterrence of general, or all-out nuclear war. While this is not the most likely kind of conflict that would involve U.S. military forces, it does pose the most serious threat to our country and to civilization. Deterrence of general war is still the primary and priority responsibility of our military forces. Without this, we would not have the option of dealing successfully with lesser threats.

When we think of deterrence of general war we think of the Strategic Air Command and the Polaris fleet. These thoughts bring forth another term with which the American public is not so well acquainted — "mixed force." A continuing mix of manned and unmanned systems as far into the future as we can conceive, can continue to generate the military advantages of a mixed force. Concomitant thoughts turn to the requirement for a follow-on manned strategic system in the 1970s, specifically AMSA, the Advanced Manned Strategic Aircraft (B-1).

When the Air Force emphasizes the need for follow-on bomber aircraft to replace its present manned force, it is not downgrading the missile. In fact, the Air Force has consistently added to its missile inventory qualitatively and quantitatively. The Air Force has expended considerable talent, facilities, and money for the most advanced missions since the beginning; it has succeeded in achieving them.

The Air Force has strongly supported the requirements for AMSA by the results of detailed analytical studies and war games, evaluating AMSA penetration capability. The advantages of a mixed force are clearly shown in these studies, because of the attributes of AMSA which complement missiles.

AMSA can provide a flexibility which is essential to a controlled response in a manner different from the flexibility provided by missiles. The 1962 Cuban crisis was an example of the necessity for manned aircraft in a cold war. Timely, on-call reconnaissance provides the basis for selecting the best courses of action.

Manned reconnaissance discovered the introduction of offensive weapons into Cuba. It documented that aggression, the build-up, and subsequent withdrawal. The reconnaissance and alert status of our manned aircraft represented more than a tactical opposition to Cuba. Our President could validly state that any aggressive action by Cuba against any country in the Western Hemisphere would be considered an attack by the Soviet Union upon the United States and would be met with the full strategic retaliatory strength of our nation. Because of this credible strategic capability, our diplomats were able to negotiate, and win.

Another example of the contribution of manned bomber aircraft in the cold war is in a "Show of the Flag." Unlike missiles in a hole in the ground or under the sea, manned bombers provide visible evidence of our national intent and determination. Several years before Cuba, evidence of our determination and resolve was made during the Lebanon and Taiwan situations. Combat aircraft were deployed; our air units were fighting fit and demonstrated this fact, both to friend and foe alike.

"Limited Strategic Conflict" is a term that has received attention in Air Force thinking. It refers to the capability to deter or defeat enemy aggression through the conduct of precise and discriminating operations against selected military targets in the enemy homeland. This capability is designed to fill the gap between the full scale general nuclear war and the upper intensities of theater operations, and thus to provide more options for controlled response. In limited strategic conflict, manned strategic aircraft provide unique capabilities which cannot be duplicated by other systems. Not the least of these is the capability of the crew to observe and report immediately the results of the mission, as well as other vital real-time information on the course of the war.

A mixed force of manned aircraft and missiles provides the Commander-in-Chief with the widest range of choices in responding sensibly to any kind of international crises. In a world with changeable power and political relationships and unpredictable future events, situations, and breakthroughs, conflicts are always possible. The most versatile, usable, adaptable, discriminating, responsive, effective, circumspect, and controllable kind of strategic mixed force is required to meet the challenge and keep the peace.

It is for these reasons that the Chief of Staff, USAF, is emphasizing the requirement for an Advanced Manned Strategic Aircraft to be operational in the 1970s. It is important that the Contract Definition Phase begin as soon as the current Concept Formulation Phase is completed for AMSA to be available as soon as possible. Aircraft under study do not deter; only available, operational, alert, and effective aircraft stop a would-be aggressor. Security during the 1970s depends on the decisions we make now. It is the future that is important, and we must get there. We must win with what we have at any given time. Plans unknown to the enemy and designs on the drawing board do not deter. We must ensure the capability for winning through the 1970s and beyond. If we cannot win, we cannot deter.

The United States has been in a strong deterrent position for general war with a mixed force. As far into the future as we can see, a mixed force offers many advantages. Thus it is essential to have a follow-on Advanced Manned Strategic Aircraft. The strength to win is required, or we cannot deter. Lacking this strength would sacrifice many advantages, one of which could be victory itself with the resultant loss of our freedom.

AMSA (B-1) Penetration Capability (1966)

I was responsible for the Air Force for the Mission Analysis for AMSA (Advanced Manned Strategic Aircraft — now B-1). The studies associated with that showed the B-1 could penetrate in its era. Many studies by various groups since have confirmed that. Otherwise, the Air Force would not have it in its priority force structure program. It has never been proved that the B-1 cannot penetrate. Such assertions are based on emotion, prejudice, or whatever

AMSA — *A Most Survivable Aircraft*. Where are the studies that prove the B-1 cannot penetrate in the future?

What are the titles?

Who are the authors?

When were the studies completed?

What was the contribution of the military?

What methodology or analytical tools were used?

What is the credibility of application of those tools to the real life operational realism of a strategic campaign?

How credible was the pre-mission planning?

How appropriate were the tactics?

Is an entire campaign evaluated?

How were the individual encounters, between the penetrator and Surface to Air Missiles (SAMs) identified, defined, and evaluated?

How were the total campaign force survivability and target destruction effectiveness evaluated?

If one had the authority to define and insure the enemy threat in any given future period, determining the military force required to meet that threat would be straightforward. However, if we should tailor our strategic strength for an incorrect, or far too narrow an enemy capability, intent, or strategy, and find that in five to ten years a threat existed different from that postulated, it could prove embarrassing. Not only would we find that billions of dollars had brought us an ineffective military posture but probably little could be done, because of the long lead times required for the design, development, and production of weapon systems of the sophistication required. What we do now determines the future. We should cover several bets.

How well have we done in the past 20 years in postulating the threat 10 years into the future?

Has the threat remained static or narrowly focused over periods of 10 years?

Or has the enemy altered the threat, perhaps due to his evaluation of our changing capability or due to an increase (or his assumed increase) in his own capability?

In whose estimation should we appear strong? Our own or the enemy's?

In what areas should we be strong? Can we define the important ones exactly?

Should we expect the enemy to think as we think?

Because we think the enemy should choose a course of action, will he?

Much has been written and stated about the survivability potential of manned bombers in the future. How well mathematics can judge the various considerations of a real life strategic combat penetration situation has not been demonstrated. How well can the civilian with his computer judge the capability of a highly specialized, highly trained, and highly motivated military planning strategist and manned bomber crew? Judgment is required, as mathematics cannot handle all the facets involved.

When a manned bomber penetrates enemy defenses, there is a chain of events which must occur for any defensive system to be effective against it. Let us briefly consider the five major steps in a defensive action: Detect — Identify — Commit — Intercept — Destroy.

We never have denied, and probably never shall deny, the enemy the first two steps. The Germans watched us complete our formations over English bases. They always knew we were coming. Identification was no trouble. If the attack was at night it was British, if by day, American.

We have not yet invented or found the invisible man, or the invisible airplane. The AMSA will be detected by radar or by Infra-Red Airborne Early Warning, IR-AEW, although a low level capability puts stringent requirements on the enemy to do so. The point is, we have always been detected and identified and we probably always will. That is not to say that reasonable things should not be done to the AMSA to reduce the effectiveness of the enemy's radar and IR detection devices, both with technology and with tactics. The effectiveness of tactics should not be underrated.

There are things being done that can deny the enemy successful accomplishment of each of the last three necessary steps; Commit — Intercept — Destroy. Before proceeding, however, the point should be made that confusing, reducing, and limiting information and capability (as well as completely denying) has a tremendous pay-off to a bomber which travels at high altitude and at very low level with the speed of AMSA.

The following could be available to deny, confuse, reduce, or limit: ICBM (Intercontinental Ballistic Missile), Air to Surface Missiles (ASM), Speed/Altitude, IR flares, Decoys, Electronic Countermeasures (ECM), Bomber Defense.

The appropriate use of the above attributes is called tactics. It should be recalled that in World War II we had available only the last two and did rather well. Also, Hitler had only an area as large as Texas to defend while Russia has a perimeter of 19,000 miles.

North Vietnam had the most sophisticated and concentrated defenses the world has ever known. Yet we penetrated with B-52s. Though some were lost during Linebacker II, the loss rate was not excessive. The figure below shows the problems for the Soviets. An outline of North Vietnam is overlaid on the USSR. Both, of course, are to the same scale.

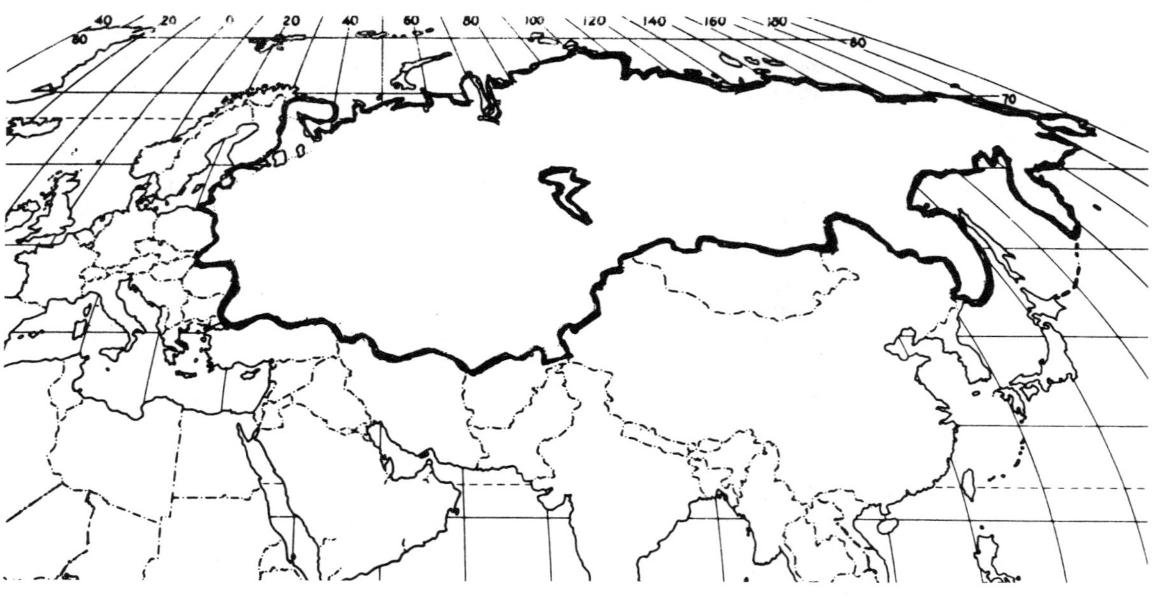

The manned bomber defense problem of the Soviets stands out. Since manned bombers penetrated the sophisticated and concentrated defenses of North Vietnam, what would the Soviets have to do to get even as good a manned bomber defense as North Vietnam, which our bombers penetrated?

It is a technical, economic, and real-life impossibility.

Another point to be remembered is that the offense can practice in detail over and over exactly what will be done in war, whereas the defense can only guess as to the enemy's tactics and set up exercises accordingly. One should extrapolate carefully to enemy defensive capability from defense lessons learned in exercises with friendly forces simulating defense.

Also, historically the first attack was most successful since it confused the defenses more than subsequent attacks. The next war will not be a war of attrition as was World War II, and thus permit the defense to benefit from its learning curve.

Today I see certain persons who are prone to assume that if the enemy has, or can get, a defense that can operate in the same environment as the offense, that therefore the offense can't live. History will not bear them out. Simply because the Cleveland professional football team knows that their opponents have players who can tackle is not sufficient reason to prohibit Jim Brown from running against them. And they still recruit offensive backs. And they still score touchdowns.

Many consider the surface-to-air missile a new dimension. It is indeed a new threat and SAMs will shoot down some bombers just as the fighters will. But it is only another way to provide defense, and an alert, modern, well-coordinated total offensive thrust still provides a threat with which the ground environment cannot completely cope.

At best, it's difficult to evaluate attrition. Mathematics and machines cannot take unrealistic and wrong assumptions and give correct answers. Many attrition studies have been made, but if the air crews should happen to act like trained military men rather than like the attrition computer, the results could be grossly different. In other words, the attrition computer has great difficulty in properly handling military judgment and experience, which can be described in one word — tactics.

Prior to World War II a Dr. Ludwig, a major in the German Air Force, assisted in the evaluation of the effectiveness of German anti-aircraft artillery. The theoretical results came out with a 0.25 per round kill probability. Some actual tests against aircraft confirmed this figure. Hermann Goering believed this figure. This is why he said that if Berlin were ever bombed, his name would be Myer. It's possible Hitler believed this and it could have been an important input to his decision to start the war. Why was the actual effectiveness only one twelve-hundreths of this, with the total World War II attrition to bombers by fighters as well as ack-ack only 1.8%? Theoretical studies suffer since the computing machine can do only what the human tells it to do. Further, the German test results confirmed again that the experience of one aircraft flying against one ground defense unit means nothing. The interaction of all offensive forces against the total ground environment must be evaluated. Obviously, this cannot be realistically tested, except in actual combat.

During World War II General LeMay's ack-ack experts told him it would be suicide to attack Tokyo at 5,000 feet. Yet one of the most destructive raids was accomplished at that altitude with no more losses than usual. Why? A trained military strategist can judge certain situations better than mathematics can.

The Aberdeen Proving Ground had made certain calculations of the survivability of B-17s and B-24s against German fighters. After the war they obtained considerable German fighter combat film from a Dr. Schmidt. A careful analysis of this film caused them to reduce their theoretical attrition results by two-thirds.

It is not expected that bomber attrition rates in a nuclear war would be as low as in World War II. But in trying to predict bomber attrition rates, one of the most important contributing factors is frequently overlooked, namely, the unpredictable factor of tactics. The commander of the offense and his tactics, more than anything else, determine the losses in any offensive action.

There are several other factors which should keep future attrition rates within acceptable limits despite continued improvement in defenses. As these defenses become more sophisticated, they must rely to an increasing degree on electronic systems which in turn are susceptible to Electronic countermeasures, ECM. Moreover, it should be remembered that attrition applies not only to the offensive, but also to the defensive.

Throughout World War II, bombers generally disregarded the defenses because they were not considered worth attacking. However, in a nuclear war the enemy's aerial defense system would become a priority target and would be attacked with one of the most effective countermeasures, namely, the hydrogen bomb. Advance attack by ICBM's and ASM's as well as each successfully dropped bomb from a bomber would take out the defenses in a wide area and permit even deeper penetration for successive bombers. The centralized netted ground environment with its vast communications, which is required to cope with the AMSA, is especially vulnerable. Further, there is an unmeasurable degradation in defense environment due to the unpredictable fog of combat operations and the destruction of ground defensive equipments, neither of which can be measured in peacetime. Consistently, history has shown that bombers did better than prior estimates indicated.

Cost-effectiveness studies have been much used, abused, and misused. However, at a constant budget level studies have consistently shown that a mixture of weapon systems provides greater effectiveness than a pure inventory of any one component weapon system at the same budget level. The extent to which this is true depends upon the enemy defense postulated. The threat (including enemy defenses) must be narrowly focused and absolutely correctly postulated, in order for any single strategic weapon system to be better from a cost-effectiveness viewpoint than a mixture of weapon systems.

The Honorable Charles J. Hitch has repeatedly pointed out the importance of time in cost-effectiveness studies, which he called cost-effectiveness-time considerations. Lower cost for a system is false economy if the system becomes obsolete, or insufficient to do the job required in some future time period. This could be caused by delays in getting the system operational, changes in the enemy threat, or an initial incorrect assessment of the threat. If the operational system is delayed for any reason, its effectiveness, and hence its cost effectiveness, could be greatly reduced. If costs become higher than estimated (e.g., more funds allocated to ensure getting the system on time because of difficulties encountered), the cost effectiveness would be reduced for the same effectiveness. Hence the least, or the most, expensive system becomes degraded or worthless in the future if the threat does not exist or has changed significantly from that for which it was designed. Conclusions:

> Our military strategy should provide many and different problems to the enemy even though he may have the capability of countering them. Building and maintaining defenses against a very low level bomber threat, a high speed bomber threat at high altitude, and against ballistic missiles could tax Soviet resources.
>
> One cannot categorically state that the B-1 cannot penetrate in its time period even against sophisticated defenses. It remains for the defense to prove under realistic operational conditions that the B-1 cannot penetrate.
>
> All conceivable situations where the B-1 can be extremely important do not involve highly sophisticated defenses.
>
> We should take the positive approach and logically and convincingly show that with the B-1 the real problem is not whether it can survive but rather whether defenses can hope to inflict prohibitive losses. If we have the case I'm certain we do, we should put the defenses on the defensive.
>
> We must respond quickly and effectively to such sweepiing unsubstantiated generalities as given by Col. J. H. Reinburg, USMCR, in "Survival in the Skies" in the May-June 1966 *Ordnance*, in which he states: "The bomber situation is different. In all cases where the other side can mount sizable fighter opposition bombers fall out of the sky almost as fast as attacked by fighters." He should look at history, as for example Korea, where by aircraft-type B-29s were second in shooting down the most MIGs.
>
> Mixed manned bomber and missile forces have cost-effective and flexibility advantages over a pure force.
>
> B-1 attributes complement missiles. The B-1 can
>
> - Be recalled, diverted, or retargeted
> - Gather reconnaissance
> - Employ minimum firepower with precision
> - Vary directions, patterns, and tactics of attack
> - Be secured in some periods of tension because of its combined loiter and mission capability
> - Provide time for national decision-making
> - Provide a time band during which strikes can be ordered according to the tension or war situations
> - Analyze and circumvent defenses
> - Provide a threat from any direction and thus complicate the enemy defense problem
> - Cause the enemy to allocate money and troops for defenses against it that thus would not be available for other military missions
> - Deliver with precision both nuclear and conventional weapons
> - Provide real-time strike and other intelligence reporting
> - Be visible evidence of our national resolve
>
> Manned bombers under study do not deter. Only available, operational, alert, and effective aircraft stop a would-be aggressor.
>
> Today's world requires stringent attention to economics. However, we should not be obsessed with frugality and economy to the extent that it results in strategic force insufficiency.

Command and Control (1960)

[Regardless of the type, amount, or sophistication of our military forces, both offensive and defensive, Command and Control is a very serious problem. While Director of the USAF Mission Analysis for CONUS Air Defense, I insisted that this complex problem be called C³S, for Command, Control, Communications, and Surveillance. This was to emphasize the necessary integration of surveillance with C³. Many had tended to treat C³ and Surveillance separately and in isolation, one from the other. This simply does not make sense.

The Navy has a serious problem with C³ for nuclear subs. Even if they can find some real estate upon which they will be permitted to put their ground equipment for SEAFARER (formerly SANGUINE, the nuclear sub C³ system) many significant operational problems remain.

To attempt to convince the Air Force that serious problems existed in Command and Control development and operations, I gave a talk in 1960 on that subject to the Air Force Scientific Advisory Board meeting at Medford, Massachusetts. Excerpts from that speech follow.]

The evolution of command and control systems has been interesting, if not orderly. As aircraft flew higher, faster, and farther, more sophistication was required for command and control. So a pattern was established. When it was assured that missiles would become a reality, it was only natural that command and control considerations would follow the same pattern of sophistication. Also, we have become so enamored of the capabilities of electronic developments that we never use a finger for an operation that can be accomplished by a thousand transistors.

We were late in concerning ourselves about the true military requirements for command and control of static missiles, for example, and rushed into the design of systems, with frills and gold plating, that have guaranteed a complexity and cost that may force a choice between buying the weapon systems *or* the means to control them. Is this complexity necessary for command and control of earth-launched missiles? And how does the situation change for space systems?

Though I can't offer any solutions, I will talk a bit of some of the things that may merit careful consideration in striving for solutions. First of all, the military must lay out the military characteristics and military requirements. Then industry should tell the military how it can meet those requirements. Our usual approach has been to find out from industry what is possible, and then mold our needs to that, rather than vice versa. If a contract to determine what is possible is awarded to a concern that likes to make black boxes, you will inevitably end up with a lot of black boxes. What else these black boxes can do, besides that which is required, will amaze you even though it may not help you.

Consideration of command and control would be incomplete without attention to how a war might start. The military must make every contribution it can to help reduce the time required for national decision-making in any emergency. Lack of success in this endeavor could negate some or all of the capabilities achieved in command and control systems.

Also, it is interesting to reflect on how the war might end. Can command and control be assured, in most imaginable devastating situations in the world after a nuclear exchange, so that one side can communicate surrender and the other can guarantee no further destruction? Some have suggested the possibility that we should spare the enemy's command and control, so that we can negotiate with him at any phase of the encounter. Others have maintained that when the next Hitler destroys himself, there will always be an admiral running around somewhere who will be willing to surrender. Too often, unfortunately, we can think only of the extremes in a given situation.

The most important single attribute of a command and control system is to be available when it is most needed, and in a condition to function effectively. Thus, command and control systems must be designed for war, but with the capability to operate for training and other purposes in peace.

Now I wish to speak of operational evaluation — that is, estimating, as accurately as possible, the operational worth of the system in helping the force to carry out its wartime mission. It is certain that the true, ultimate effectiveness of command and control will be proved only by combat use. We cannot, however, wait for the war to find whether we have effective systems. Operational evaluation must be accomplished during peace and, to be useful, must simulate combat operations realistically.

I am saying here that operational evaluation of the system is a very important activity, and it is not too extreme to say that our national survival might depend upon its being done properly.

Chapter 14
Public Media

Some Associations

One of my most difficult chores during the 13 years I was with Hq SAC, was dealing with representatives of the public media. Sometimes I would refuse for months to see any of them. Then on occasion I would go on speaking "binges." One year I made 24 major addresses from coast to coast. I would spend much time in preparation, so that I could speak "extemporaneously," rather than reading a speech. It also was necessary to sort out the material that was for public release from that which still had some kind of restriction on its release. With so much information from so many sources, it was difficult to remember the source or the present status of the information. I would write questions and give them to the Public Information Officer (PIO), requesting answers in writing. Then I would study them, to know what I could say about a given subject.

Once the *Christian Science Monitor* had a nearly full-page special about me and what I said about SAC.

I had trouble avoiding the press on trips to Europe. Twice I personally heard on the radio that I was visiting — once in England and once in Germany (I already knew I was there). I was in Germany shortly after SPUTNIK I was launched, and it was difficult to avoid discussion of the implications.

Jacqueline Cochran was asked to tape an interview for future showing on TV in Omaha. She said she would if I participated. I obliged.

I was asked to give a luncheon address to the Chamber of Commerce in Des Moines, Iowa. I arrived at my hotel about midnight, from Washington, D.C., and found that a press conference had been scheduled for 8 the next morning. One of the questions was when the Missouri Valley was going to get ICBMs. Because of my prior homework I was prepared. I replied that I knew nothing about Des Moines but the Omaha complex was scheduled to become operational in 1959. The next edition of the Des Moines *Register* reported that I had said that the Missouri Valley was going to have ICBMs in 1959.

(I was concerned that the governor, an Iowa senator, or the representative would check with Washington and ask why they did not have such information. But I heard nothing more from Iowa, nor did anyone. However, my home newspaper, the Omaha *World Herald*, got mad at me for scooping them.)

I also found that I had been scheduled to appear live on two different TV stations during that Des Moines visit. I did that, and gave my luncheon address.

Joseph Alsop

Once Joseph Alsop visited SAC Hq for two days. He had already been at one of our bases and on a training mission. I knew that Joe had the reputation of speaking only to God. Somehow it became my lot to squire him around. When I saw General Power at noon I said that I must be too good an informer. I not only had not yet been insulted by Joe, but after the first half-hour I was "Zim." At his insistence, I was Joe's guest at dinner that night.

I was quite concerned that some things would be reported out of context. Sure enough, two days after Joe returned to Washington, I got a message that I should phone him. A fantastic thing happened: Joe asked me to forget something *he* had said. His several articles about SAC were good and factual.

A good way to get into trouble is to deal with representatives of the public media. I include some members of Congress in this category.

The Miamisburg (Ohio) News, 11 September 1958

> The Miami Valley area has had more than a passing interest in the Strategic Air Command, the Nation's number one defense organization, due to the fact that a native Miami Valley man, Carroll L. Zimmerman, holds down one of the higher posts in the SAC organization.
>
> Mr. Zimmerman, son of Louis A. Zimmerman, 802 North Fourth street, was born in Miamisburg, graduated from Miamisburg schools, and taught physics in college before he began his climb into the higher echelon of the Strategic Air Command. He now holds the post of Chief Scientist, SAC.
>
> Since it was established, and until recently, little was known of SAC and its operational field, but a new policy now encourages the dissemination of as much

information as possible about SAC to the general public. SAC has come around to the belief that the United States will be stronger if private citizens are sufficiently aware of their role in maintaining a critical margin of deterrence. Toward this end persons high in the SAC organization are not only making themselves available for press interviews, but are making frequent speaking appearances throughout the country.

A recent issue of the *Christian Science Monitor*, published in Boston, carried an interview of General Thomas S. Power, now SAC Commander-in-Chief, and Carroll L. Zimmerman, scientific advisor, parts of which may be of general interest locally.

The *Monitor* article said in part: "Interviews with SAC's top command yield surprisingly simple answers to the big question about America's security and survival:

" 'Prospects are excellent. The risk of getting into global war is small,' states Mr. Zimmerman — 'provided Americans as individuals face and handle the dangers.'

"There is no single period of greatest peril. But as the Soviets gain greater and greater capability through their new missiles and other weapons, the danger will increase proportionately.

"There is no question but that the mounting threat can be met, and SAC's programs are geared to meet it.

"But the variable, the unknown, the most important factor, is the individual American. There is currently a graver danger in failing to have the public informed than in the military power of the enemy itself.

"SAC is confident in its own operational abilities, its capacity to maintain a global nuclear striking force in unending around-the-clock readiness . . . its more than 70 U.S. and overseas bases . . . its near-sonic jet bombers of today and its supersonic bombers of the future . . . its eventual missile forces carefully time-phased and blended with bomber forces to assure no gap in our defense posture . . . its 15-minute alert system with (as an ultimate goal) one-third of the bomber force on perpetual alert."

In the interview General Power was quoted as saying: "We can hope to deter limited wars only if we convince the Soviets or their limited war proxies that we are determined and ready to employ our best weapons, including SAC, to quickly stop and decisively win any limited war that threatens the peace of the world. Our determination to take decisive action in limited wars need not be weakened by fear that such action might expand a local war into a general war with the Soviets. For the very moment our country gets involved in a limited war — either directly or indirectly — SAC goes from a partial to a full alert status, depriving the Soviets of their most important advantage — surprise."

Mr. Zimmerman added: "The Russians would never start a local war with the thought in mind that it would become a general war. Nothing can guarantee 100 percent security; we have to take some risks."

He went on to say that the job of the American people is to make their intentions known and clear. One part of that job, Mr. Zimmerman urges, is to insist that a nuclear test moratorium must not be declared. "Those who advocate such a moratorium say that we are ahead at this point. But if the Soviets were really behind, do you think they would agree to a moratorium?"

What he thinks is the futility of a nuclear ban is summarized by General Power: "Nuclear weapons offer such decisive advantages to an aggressor that anyone who is unscrupulous enough to attack a peaceful country could hardly be expected to honor agreements which would make the success of his attack doubtful, if not impossible."

In another article which appeared in an August issue of the Omaha *World-Herald*, Mr. Zimmerman was quoted as saying, "There must never be a war of nuclear weapons. Nobody wins it!"

He went on to tell why, as follows:

"There can be no effective defense against intercontinental ballistic missiles. We would have to knock down 100% of them. An enemy could launch one thousand missiles. Even if the attrition rate could be 95% — and this is most unrealistic — 50 missiles would get through. The toll from 50 missiles carrying thermonuclear warheads would be fearful. If 50% of the missiles could be stopped, five hundred would get through. This is the situation facing the United States."

As a result, Mr. Zimmerman said, SAC is "doing everything possible to take the profit out of war for an aggressor." Though SAC "will never strike the first blow" it is ready to retaliate instantly, he declared.

Mr. Zimmerman said he does not believe SAC will ever be a total missile force. Manned bombers can be in the air — "and many are in the air at all times" — and can be diverted to a real target in the event of an attack. With missiles this is impossible. "You don't recall missiles once they are on their way," the scientist said.

Mr. Zimmerman said the Atlas missiles, soon to be Omaha based, will be the basic missile weapon for at least ten years. "And I hope they sit there forever!" he declared.

Chapter 15
Individuals

General William H. (Butch) Blanchard

When Butch Blanchard was Vice Chief of Staff, USAF, and I was with the aerospace industry in California, he phoned to ask that I see him the next time I was in Washington. I did. he strongly recommended that, since I knew General LeMay better than anyone (those were his words), I write a book about him. He said it would have more impact written by a civilian than by a military man. He said he would open the Air Force Archives to me, and would personally contact selected individuals to send me stories, anecdotes, and so forth. Butch was dead within a week from a massive heart attack, so the project was never launched.

Charles A. Lindbergh
(Written on 27 August 1974)

It has been my good fortune to have had an intimate association with many world famous personages — truly great men. I have not talked or written much about those associations because they were mostly confidential exchanges between friends. The Charles Lindbergh I knew, like General Curtis LeMay and Professor Johnny von Neuman, was a very different person from the Lindbergh that the press and the public knew.

His death yesterday obliges me to record some of the things I know about him.

I first met Charles Lindbergh in 1952 on the 25th anniversary of his famous flight. I suggested that others, no doubt competent, had failed. Six died in their attempts. Why was he successful? His answer was immediate and forthright: his sponsors gave him the money and didn't compromise his chances for success by interference, as the sponsors of the other attempts had.

For six years I spent three or four days at least twice a year with Charles. Occasionally we were roommates, but usually we had adjacent hotel rooms. He didn't particularly like to drive, so I would rent a car and drive, and he would navigate. Occasionally, I would make some choice comment about the quality of the navigation available to me. Remember that, without a radio (too heavy), he made landfall at Ireland within six miles of his planned point of entry.

Charles told me many stories about barnstorming, acrobatics, wing-walking, mail-flying. One was about the early pilot (his name escapes me now) who took six or eight people for hire on a trip. At the destination the pilot was surprised when he walked back into the cabin. No one moved. Then the terrible realization hit him — all were dead from carbon monoxide poisoning.

Once for hire Charles took two couples on a flight. Beaucoup booze. They would throw empty bottles out of the aircraft. They played strip poker. When a piece of clothing was removed it was thrown out of the aircraft. His discussion was quite amusing of the alternatives he considered, upon landing, to get a soused, completely nude woman from the aircraft in a way that would not get him in trouble, and to a place where she would not be molested.

Charles didn't drink or smoke. He never swore. He was most considerate of others, except for newspaper reporters, whom he hated with a vengeance. He blamed them for the death of his son. Though he talked often about his son, Jan, he never said a word to me about the 1932 kidnapping. Neither did he say anything about his experiences in the Pacific during World War II.

I was much amused when he was talking about his competitors, and their aircraft, in being the first to fly the Atlantic. Without a pause, or the batting of the eye, he described the color of one aircraft as "titty nipple pink."

Though his aircraft was under guard in a hangar at Roosevelt Field, somehow two girls got in and were observing the aircraft. They saw a tube and a funnel immediately in front of the pilot's position, but considerably forward of the pilot's seat. The reason for this tube is given in the next paragraph. As Charles approached them undetected from the rear, apparently thinking it was a relief tube, he heard one girl say, "Boy! I'd like to meet him!"

The chief engineer at Ryan Aircraft in San Diego, who designed and built, with Lindbergh, the "Spirit of St. Louis," would not speak to Charles for years after the flight. Unknown to him, Charles had installed an extra 40-gallon fuel tank. The tube of the previous paragraph was fastened to it for filling. The engineer never did completely forgive Charles for possibly compromising his design and balance in a way that, in his view, could have been fatal. Thus Charles had enough fuel to go on to Rome. He briefly considered this. But when he saw the lights at Le Bourget, he decided to land. He was bitter about the vandalism of souvenir hunters at Le Bourget.

He stayed at the residence of the U.S. Ambassador while in Paris. He and the "Spirit of St. Louis" returned to the U.S. on a U.S. Navy ship.

It has been written that the night before his flight he did not go to bed, to practice staying awake. It is true that he did not go to bed, but that is not quite the reason. He had been assured that weather precluded any opportunities to take off within the next 24 hours. So he went to a Broadway show and took a show girl out afterwards. He got to his hotel room, alone, in the early morning hours. The phone rang about 15 minutes later. He was told that there was a chance for a break in the weather. He left immediately for Roosevelt Field.

When I lived in Omaha, he and I were together in Washington, D.C.; I told him I wanted to get an electric razor I had seen on sale. He said he would never have one — they were much too heavy. I told him that he and I had been brought up differently. On my first flight to Europe I did not get $1,000 for each pound over and above that required for flight safety and success.

His wife owned a chalet in Switzerland where they would go to write. When Charles found that Valerie, my daughter, could recite one of Anne Spencer Morrow's poems, he was much impressed and told his wife. Valerie can recite it today.

After World War II he was on a trip through Turkey alone. He had rented a Volkswagen, and was stopped looking at a map. A gendarme, or whatever they call them in Turkey, came up to him. The policeman did not speak or understand English. But he understood the words, "Charles Lindbergh." He arrested the imposter. It took some doing to untangle that one.

As you can imagine, he received much mail. He opened none of it. The only mail he opened was that addressed in a specific fashion to Darien, Connecticut. He gave this correct address only to a few people. I was fortunate to be one of them. He saved all his mail, unopened, in cartons. He gave the collection to the Yale University library. They informed him that if he could sign his name eight times a minute, eight hours a day, five days a week, it would require eight years just to sign replies.

I spoke of his consideration. One example: prior to his last landing at his beloved place in Hawaii, and just before he died, the pilot asked if he should circle for Charles to have a look. Charles said that he should not, as it might inconvenience someone.

His death yesterday resulted in my pilgrimage to the Smithsonian Institution, where I took another look at the "Spirit of St. Louis" and spent several minutes in silent prayer.

I cherish my copy of the book *The Spirit of St. Louis*, which he personally gave me after writing some comments and autographing it for me. This was at his own initiative and not because of any request from me.

* * *

Switzerland
Aug. 21, 1964

Dear Zim:

I start with an apology for the lateness of this reply to your July 9th letter. The letter arrived after I had left the United States, and I have just found it in a stack of forwarded mail here in my wife's chalet in Switzerland. I have been travelling a great deal and am, as usual, hopelessly behind with correspondence.

Of course I would like to see you at any time, just from a personal standpoint, if no other. I have missed the contacts we had in Washington and on Committee tours. I will be back in the United States in September, but I am not yet sure of the date.

Please don't make any change in your plans, because my schedule is too uncertain. But if you are coming east after about the middle of September, and if I can be back as I now plan, I would be glad to meet you in New York. In such case, I suggest that you let me know when you expect to be in New York and address the envelope as follows:

P.O. Box 1147
Darien Connecticut

But don't put my name on the envelope because then it will get in the general mail, and that often gets to be more than I can handle. (I don't keep a permanent secretary.)

With best wishes always,
/s/ Charles
Charles A. Lindbergh

Col. E. F. Calafato, USAF Ret.

At SAC Hq in Omaha I decided in 1953 that I needed a young executive officer. I wrote the attributes I thought I wanted and took them to the director of Personnel, Col. A. J. Russell. (Later Colonel Russell retired with three stars, his last position being Assistant Vice Chief of Staff, USAF.) Colonel Russell had his people prepare the appropriate inputs for the computer. Six names popped out.

Capt. Eugene F. Calafato was the second candidate I interviewed. I asked Colonel Russell to cancel my interviews with the rest; I had found my man. When he reported for duty in June 1953 I already had a Gene and a Jean in the office, so I told him he was Cal. He has remained Cal to me. Captain Calafato worked with imagination, ingenuity, industry, competence, and with little supervision in helping me far and beyond those things I expected from an officer in such a position.

After a most productive and exciting tour, to me, his next assignment was to SAC's Seventh Air Division in England in June 1955. His position was Executive to the Director of Operations. Besides the overseas SAC bases there were 18 Operations Analysis offices around the world. Hence my responsibilities required frequent overseas trips, and on each trip to Europe Cal paved (red-carpeted would be more accurate) the way for me.

One time I was going from Paris to London with four-star General Lawton Collins (U.S. Army), then chairman of the NATO Standing Group. Cal had arranged for a car to meet me. When he had phoned the airport to make certain that I would be properly cared for, the WAF officer from the Protocol Office, with whom he usually did business, did not answer. A man did. Cal explained that I was coming in, how and when, and the arrangements that had been made. He wanted nothing to go wrong. After all the details were carefully explained, as was his wont, he asked to whom he was speaking, so that if something got fouled up, he would know who to have hanged from the lamppost.

The reply: "Captain Calafato, this is Major General Wilson. I guarantee you that I can take care of Mr. Zimmerman."

Cal was living in a house on the British economy. During one visit I was his house guest. One night we went next door (Cal, his wife, Edie, and I) to his British neighbor. I like to dance and girls like to dance with me — even today young lovelies ask me to dance. I've never found a female that could dance me under the table, but I almost did in Cal's neighbor. We danced and danced (at her insistence), and so did others. I had long since taken off my jacket, and my host did not throw me out. Even a staid British gentleman can break inbred tradition. Soon after I took off my jacket, my host took his off, as well as his waistcoat (vest). Then he put his jacket back on.

In 1959, when I received the appointment of Chief, Operations Analysis, USAF, Major Calafato had just reported to Plans, USAF, from the Seventh Air Division. It was no problem getting him reassigned as executive officer to me. Besides keeping me and Operations Analysis out of trouble, he did much for our progress, and thus for the Air Force. Extracurricularly, he wrote a speech on Foreign Policy Perspectives and gave it to the World Politics Seminar at Oxford University, England. I received for transmittal to him an exceptionally fine letter of commendation from the seminar chairman.

I cannot recount all his accomplishments.

During this time the USAF carefully selected six officers to serve as escorts for dignitaries from abroad visiting USAF installations in this country and overseas. Four were colonels, one was a lieutenant colonel, and a major — Calafato. They also served as escorts for U.S. dignitaries visiting USAF installations in this country and overseas. One senator always asked for Major Calafato by name when making such trips.

The wife of a member of my office was active in the International Girl Scouts organization. When she was at a meeting of the organization in Monrovia, Liberia, she suffered a stroke. Commercial airlines and ship lines refused to accept her for passage to anywhere. Major Calafato secured the concurrence of the Commander, USAFE, the Assistant Vice Chief of Staff USAF, and the Director of Transportation USAF, for an Air Force medical evacuation aircraft to take her to Wiesbaden, Germany, and thence to Washington. I sincerely believe that no one else could have effected those arrangements. Newspapers labeled it the longest mercy flight in history.

At 9 a.m. one Saturday I received the information that Tex Alexander, the man I had sent to be Chief, Operations Analysis in Hawaii, had died of a heart attack. By midnight Major Calafato arrived in Hawaii to assist Alexander's widow with the planning needed. She certainly needed help and compassion.

In June 1965 he was assigned as military assistant to the Chief, Operations Analysis, USAF, so when I was selected to become Chief, again I inherited Cal. Besides his stellar work for me he continued to provide staff assistance to the Assistant Secretary, International Security Affairs, Department of Defense, on U.S.-Panamanian relations, counterinsurgency, and Mutually Balanced Force Reduction (MBFR). I received an unusually fine letter of commendation for his work with them, with sincere appreciation.

In June 1972 Cal was assigned, as a colonel, to SHAPE as U.S. National Military Representative, a two-star billet.

Truly he was outstanding in his dedication, contributions, and loyalty to the USAF. He was one of the most unusual, competent, gifted, hard-working, patriotic, effective, and productive men I have ever known.

Secretary of Defense Robert S. McNamara

It is an understatement that Mr. Robert McNamara was not a popular Secretary of Defense in many quarters. The reasons include his egotism and arrogance; his belief that everything can be represented by a mathematical equation; and his absolute faith in his young, often naive, and inexperienced Whiz Kids. Though some were good analysts (and I shall modify that later) they had led a discouragingly sheltered life as far as military operations were concerned.

Most Whiz Kids believed that military inputs were the least important in any assessment, because everything is a mathematical equation. Also, unfortunately, too many of them demonstrated an amazing lack of tact, consideration, and decency in dealing with the military. The SAC Adage was truth to the Whiz Kids on almost any subject. The SAC Adage: The expertness on SAC of an individual is directly proportional to the distance that individual is from SAC (or for the Whiz Kids from any subject under consideration). Those who are farthest away are the greatest experts. Those responsible for the operation are at a zero distance from it. Therefore they have zero expertness.

Some of the most obvious advocates of the SAC Adage believed strongly that SAC had an NIH (Not Invented Here) factor that seemed to prevail in the formative years. To many in the Air Staff, USAF, during the early 1950s SAC gave the impression of having a corner on brains. This supposed attitude tended to isolate SAC to some extent.

That impression is a fine example of the SAC Adage. Anyone who really knew General LeMay was well aware that he believed no one had a corner on brains. He did not even consider strange letters as "crank" until they were evaluated. He did not want to miss any possible idea. But he regarded those who had never been to SAC, yet were making decisions affecting SAC, as pseudo-experts, and he regretted that those who visited SAC for half a day or a day became "experts" on SAC. Any professional would support his own ideas against naive amateurs when the subject matter under consideration was his responsibility.

General LeMay released some information to the Rand Corporation that he refused to release to Roy Brothers, then Chief, Operations Analysis, USAF. The reason was that Rand sent a professional who spent months with us. Roy sent only a written request, without any Operations Analysis, USAF, participation at SAC.

Anyone who believed that SAC had an NIH attitude simply did not know the facts.

Though many of the Whiz Kids were junior officers, they never wore a uniform. I overheard a young Army lieutenant, who was working on Air Force problems, tell a colleague why he never wore a uniform: a young first lieutenant would have problems with senior officers, but as Dr. Joe Blow from the office of the omnipotent Mr. McNamara he could, and would, expect awesome respect from senior officers.

A study by another young Army first lieutenant almost eliminated incentive, or flight, pay for the Air Force, Navy, and Marine flight crews. Though causing quite a stir, and much effort to provide illumination on the true facts of the situation, he did not prevail.

Though some of the Whiz Kids were good analysts, most lacked experience; the ability to assess the completeness, validity, and relevance of input data; an awareness of the proper tool for the job (an expert on linear programming tended to use that tool for every job; one usually does not take a hammer to cut a board into two pieces); a capacity for astute and unbiased interpretation of the results.

McNamara had been on duty two, perhaps three, weeks when he asked for a discussion (not briefing) with General LeMay on the B-70. General LeMay took Lt. Gen. Doc Strother, then Deputy Chief of Staff, Operations, and me with him. Mr. McNamara was alone. General LeMay referred one question to me. (I've tried to reconstruct what it was, and I have failed.) After I answered, Mr. McNamara's reply was, "Yes, provided we need manned bombers."

After the meeting I commented to General LeMay that we were going to have trouble with that guy, and he replied, "If he is a big enough man for the job, we'll do O.K." The Air Force did not do O.K. The B, RS-70 was canceled.

Later I was in a much larger meeting with McNamara after his first trip to North American Aviation. He said derisively several times, "And they are talking in lots of one hundred." McNamara was right in this criticism: it was absurd for the contractor to talk that way, even though financial types under McNamara demanded that information.

At that time a young colonel was Hq USAF action officer on the B, RS-70. He did a magnificent job against impossible odds. The young colonel's name was David C. Jones.

The B-1 met a similar fate from President Carter. In my view the decision to cancel the B-1 was wrong, and potentially dangerous to the security of our nation. I really burned when a member of President Carter's staff asked publicly, and proudly (and in my estimation naively and irresponsibly), "What President would have the courage to make such a decision without inputs from the military?"

As for the follow-on manned bomber to the B-52 (the *youngest* one will be 29 years old in 1990), we should go all out to get some. Then we can convince anyone we must, how many we need.

McNamara recommended to two Presidents that General LeMay be fired. President Kennedy did not accept the recommendation. With President Johnson, McNamara's approach was to recommend against General LeMay for a second two-year term as Chief of Staff, USAF. One evening at about six o'clock President Johnson called General LeMay to the White House. The President gave him the news, and told him that he did not approve and that LeMay would be appointed to a second term.

President Johnson had neglected to tell Air Force Secretary Zuckert and Mr. McNamara of his decision. The next morning General LeMay broke the news to each of them.

McNamara's cancelation of the Air Force SKYBOLT missile was purely political, and not because of technical difficulties, as McNamara told the public. In my files is a written statement to that effect by someone who was in the meeting with McNamara and the President when that decision was made.

Even President Johnson decided he must get rid of McNamara. He was kicked (upstairs? downstairs?) to the World Bank as president. He retired from that position in 1981.

The Gaither Committee

In the summer of 1957, after General LeMay had become Vice Chief of Staff, USAF, and General Power had become CINCSAC, President Eisenhower established the Gaither Committee to advise him on Civil Defense. Through channels, the committee requested a SAC representative. Having been advised by Hq USAF that the committee's focus was on pouring concrete for underground shelters for the U.S. population, the SAC staff supported General Power's inclination to shrug off the request — "the press of SAC business, you know."

Additional pressure was brought to bear (after all, the President was on their side), and General Power agreed, reluctantly, to assign Stan Lawwill. In the meantime, arrangements had been made for the committee to be briefed by both Hq Air Defense Command and Hq SAC. Lawwill joined the committee on 4 August 1957 at Colorado Springs for the Air Defense Command briefing, and returned with them to Offutt for the SAC briefing.

As might be expected, there existed a full spectrum of "SAC Briefings" ranging from a completely unclassified description of SAC's mission, facilities, and capabilities (which could be delivered to any member of the local Chamber of Commerce) to a highly classified disclosure of the current SAC War Plan, based on tested capabilities of weapons, crews, and units and the most sensitive intelligence information available (today called "Sensitive Compartmented Intelligence," in those days called "Green Door" because the special, deep-underground intelligence area at SAC Hq was entered through a green door). The briefing given the Gaither Committee by SAC was nearly at the first-mentioned end of the spectrum. After all, SAC had little interest in civil defense shelters: SAC's assigned mission was to deter nuclear war and, if that failed, to execute the War Plan in optimum fashion.

The committee members recognized that they had received the "idiot treatment" at SAC and on the flight back to Washington discussed bitterly among themselves the difficulty of doing their assigned job if they couldn't have access to necessary military information.

On the trip to Washington Lawwill was advised of the committee organization — advisory panels, four working groups including the "Evaluation and Quantitative Assessment Working Group" of which Lawwill was Deputy, and a steering committee consisting of Bob (R. C.) Sprague, Bill (S. C.) Foster, Jim (J. P. III) Baxter, and the heads or deputies of the four working groups. In the course of this orientation, Lawwill recognized that the Air Force had grossly misjudged the function of the Gaither Committee. (The point has been made elsewhere that an important function of Operations Analysis was finding out what the problem *really* was — in contrast to what the client had said it was.)

The next day Lawwill called the director of Operations, Hq USAF, then Maj. Gen. Wheless. ("Shorty" in his many years at SAC had interacted frequently with Operations Analysis.) Stan received an immediate audience and explained the purpose of the Gaither Committee, or, as it was already being called, the Security Resources Panel of the Science Advisory Committee to the President. President Eisenhower had been briefed early in 1957 on the recommendations of the Federal Civil Defense Administration — in essence, that the United States must spend about $40 billion, primarily on fall-out shelters, to protect citizens in the event of a nuclear war. President Eisenhower's reaction was interpreted to be, "I don't doubt that $40 billion spent on fall-out shelters would save many Americans. However, I wonder whether there is a better way to spend $40 billion to save Americans from death in a nuclear war." The President asked, through the National Security Council, that the Science Advisory Committee establish a high-level group to advise him on this matter. Hence, the Gaither Committee.

General Wheless immediately buzzed General LeMay on the squawk-box saying, "Boss, I was wrong about the Gaither Committee. When can you see Lawwill to get the straight scoop?" General LeMay's "Now" had Lawwill on his way to the front office. After hearing Stan out, General LeMay called in Lt. Gen. Jake Smart, Assistant Vice Chief of Staff, and said, "Jake, we were stupid. See if you can square things with Sprague and Foster. Lawwill will brief you on the way to the Executive Offices of the President."

On the way Lawwill gave General Smart the important background. General Smart handled the matter with the greatest finesse and reached agreement with Sprague and Foster that General LeMay would visit, would personally brief a selected few members of the Steering Committee on Air Force plans, and would arrange special clearances so they could return to Hq SAC for a proper briefing on SAC.

Naturally, Lawwill was included in the "selected few." He reported that General LeMay was magnificent; he accepted full responsibility for the Air Force's faux pas, was completely candid in answering questions, and regained for the Air Force the respect of the panel leaders.

A special flight was set up to take the selected few to

SAC. General Power met them and apologized for SAC for the misunderstanding. He introduced the "Green Door" briefing and stayed with the group the entire day, answering questions with great precision and competence and, as requested, philosophized with clear forethought and understanding about deterrence, Soviet future capabilities and intentions, and so on. In effect, the committee was turned around and pointed toward preventing war rather than surviving war.

Over the next three months the panel received many briefings, reviewed and evaluated innumerable studies by the military and other government agencies, and began to draw up for the President its priority list of recommended expenditures. One day Sprague and Foster called Lawwill into their office and suggested that the panel report might have more impact — or less adverse comment — if Lawwill's name did not show on the steering committee. The reason, of course, was that all of the committee's top priorities for expenditures were to strengthen SAC's war-fighting, therefore deterrent, capability. Without hesitation, Stan agreed. He had long since learned that good work is recognized whether it is advertised or not.

As might be expected, careful readers of the Gaither Report noted that the transmittal letter says, ". . . the steering committee, which includes the director, the co-director, and the heads of the four working groups, have full responsibility for this report." The signature element, however, shows the list of the steering committee also containing the names of the deputy heads of the other three working groups. Since the report also shows the panel organization, including heads and deputies of the four working groups, omission of Lawwill's name from the steering committee was not very subtle.

Chapter 16

Scientists

Scientists are widely misunderstood by a fair portion of the public and the press. Most scientists are considerate and rational human beings, especially the good and great ones, but a small percentage of scientists create a negative impression for the public and the press. They appear to be more interested in getting their names often in the public media and claiming to be an expert scientist, rather than working at being one. For some of them the fact is they would fail if they tried to become one. Many of the press do not understand, and they transmit their lack of understanding to the public.

I have personally heard or observed all of the items included below except for the following:

Albert A. Michelson. The statement by him is well documented. My contribution is the observation about modern physics.

Robert A. Millikan. Told to me by Clark Millikan, his son.

H. Guyford Stever. Told to me by him.

Physician. A story I heard years ago, but I cannot recall the source.

Nightgown. Told to me by Maj. Gen. David Wade.

* * *

Albert A. Michelson became famous for his experiment that measured the velocity of light, conducted at what is now called Michelson Hall at the Naval Academy at Annapolis, Maryland. He said the following in 1899: "The more important fundamental laws and facts of physical science have all been discovered, and these are so firmly established that the possibility of their ever being supplanted in consequence of new discoveries is extremely remote. . . . Our future discoveries must be looked for in the sixth place of decimals." I find it extremely interesting that this was said several years after the discovery of X-rays (in 1895), which provided the precursor to the dramatic advances in modern physics.

* * *

Robert A. Millikan was a celebrated physicist early in this century. He is famous for his "oil drop" experiment which resulted in the first determination of the charge of an electron. He won a Nobel Prize. One day he overheard his maid on the phone: "Yessir, this is Dr. Millikan's residence. Oh nosir, he ain't the kind of a doctor that does anyone any good."

* * *

C. C. Kiess. For two years I was guest research worker at the National Bureau of Standards. This meant that I was permitted to use some of the laboratory research equipment, but I didn't get a dime for it. Dr. Kiess was my supervisor. One day a woman came collecting contributions for the Red Cross. Dr. Kiess gave her five cents. Before the startled woman could say anything, he gave her a dollar, saying "This is to hope that that five cents gets to a place where it can do some good."

* * *

Edward Teller came to this country in 1935 to the physics staff at George Washington University in Washington, D.C. At that time I went to GW as Associate in Physics to continue graduate work beyond the M. A. which I had at that time. Dr. Teller became my sponsor. Besides the courses I took from him, I gained in two years something like 16 semester hours of credit in research. Some years after I left GW we became closely associated in several activities, particularly in the early days of the nation's efforts in first atomic, and then nuclear, research.

At Teller's request I arranged for SAC crews flying at high altitudes and high latitudes to collect data for cosmic ray research for Robert Brode, his colleague at the University of California. (Incidentally, Robert Brode was one of the three male triplets to graduate from Whitman College. They were the first triplets to each receive a Ph.D.)

During the time I was with Hq SAC in Omaha I was preparing to go to an Air Force Scientific Advisory Board meeting just outside Boston. My daughter, Valerie, about 12 then, asked what I was going to do. "Give a speech."

"Who will hear it?"

"Many people, including Dr. Teller. . . ."

I was stopped short there when she exclaimed, "Do you mean to tell me that *he* can learn something from *you*?"

Years ago Edward gave a speech at the University of Missouri. He told me about how he was introduced: he had told the lad who was to introduce him that he did not like long introductions, and he didn't want to be known as the father of anything. So the lad's introduction was, "Tonight we shall hear from Dr. Edward Teller, who was born in Budapest. You can tell he's from Hungary because he looks exactly like Zsa Zsa Gabor." Then he sat down.

You may have noticed Dr. Teller's limp. While a lad in Budapest he was trying to "hook" a free ride on a moving trolley. He slipped and one foot was cut off.

* * *

George Gamow and Teller were students under Niels Bohr, the famous Danish physicist, at the same time. Gamow was on the faculty of GW and persuaded Teller to join him. On one visit of Dr. Bohr to his former students I had the opportunity to spend some time with him, including playing pingpong at the Faculty Club. Gamow loved wine, women, and prize fights. He would go to extreme efforts to see any boxing match he could.

One of the courses I took from him for a semester was the application of the special theory of relativity to quantum mechanics. The second semester I took the application of the general theory of relativity to quantum mechanics, at the same time it was being developed by him.

In class and in seminars neither Teller nor Gamow ever missed a 4-pi-C^2, but arithmetic was a gray horse of a different hue. Teller was better than Gamow. The following is typical of what happened frequently: Gamow would say, without turning aside, "How much is six times nine?" Teller would say, "Fifty."

During this time Gamow became a father for the first time. Often at the start of a lecture he would go to the blackboard and draw a graph and say, "The baby book says this." Then he would draw another graph and say, "But ours does this, so we don't know what we have." Once while the baby was still quite small he was required to fill out a form about it for some purpose. One entry was HEIGHT. He struck that out and substituted LENGTH.

After accepting an invitation to give a talk in Puerto Rico, Gamow found that it was to be given in Spanish. He proceeded to learn Spanish. And once at the sedate Cosmos Club in Washington, D.C., he gave a talk on "Cosmetic Rays." (He applied cosmic ray theory to female beautifying.)

In the beginning either one of them would occasionally be unable to think of some English word. They would say it in German, and the class would translate it into English.

Gamow spent his last years at the University of Colorado, at Boulder. That university has functioning today utilitarian monuments (that is, laboratories) to both Gamow and Teller.

Professor Birge

While I was head of the Physics Department at Whitman College, I was audacious enough to visit personally physics graduate departments of selected universities. My purpose was to see if they were good enough for my physics graduates. Professor Birge was chairman of the Department of Physics at the University of California, Berkeley, when I visited there. He received me politely for about five minutes and then called a graduate student to give me a tour. Birge's contribution to physics had been doing what Dr. Michelson had said in 1899 was the future of physics: measuring the value of physical constants to more decimal places. When we went into the cyclotron research laboratory, Professor Lawrence was there. I couldn't get away from him for two hours. Not that I wanted to, but I was apprehensive about the possibility of wasting his time. This experience is an excellent example of the Zimmerman Law.

* * *

Louis Ridenour was giving a presentation at a symposium at the radiation lab at MIT early in World War II. One projection was a picture of a German radar. It was marked British Most Secret, U.S. Secret, and German Obsolete. I am reminded that the British often have quaint ways of expressing things. Prior to and early in the war they did research in High Frequency (HF). Then they progressed to Very High Frequency (VHF), and then to Very High Frequency Indeed (VHFI).

* * *

Din Land invented the Polaroid camera and was architect, builder, owner, and manager of the Polaroid Company. Many years ago he went to his office in his laboratory and was surprised that no one was there yet. He got engrossed in something, but later when he awakened from that he realized that still no one was there. This baffled him. He phoned someone and found that it was Thanksgiving Day. Though Polaroid seems to be in a bit of financial difficulties today, and Land has given direct operating responsibilities to someone else, I make two observations: (1) Don't underestimate the effectiveness of his continued influence with the company, and (2) He is probably the wealthiest scientist in the world today.

* * *

George Kistiakosky was Scientific Advisor to President Eisenhower. Prior to that, he developed the shaped high explosive charge for the early atomic weapons. He received a Presidential citation for this. After World War II he wanted to clear some property he owned, and he wanted to use dynamite to do some of it. He found he could not purchase any because he did not have a permit for its use — it could be sold only to those that knew what they were doing. His many and lengthy attempts to get a permit and/or some dynamite are quite amusing. They all failed. Finally an idea struck him. He took his Presidential citation to the office that issued permits. He got a permit. This was one time a citation was of more use than adorning a wall. This reminds me of another example of bureaucracy. Though once I was head of a College Physics Department, I was not eligible to teach physics in any high school in the nation. I did not have the required courses in Education.

* * *

Darol Froman was associate director of the Los Alamos laboratory. Once we observed together a test of the propulsion rockets for NAVAJO. A single propulsion unit consisted of three 135,000-pound-thrust engines. When they were ignited simultaneously, it was quite a sight. After the noise subsided, Dr. Froman said to me, "Even chemical explosions can be impressive."

* * *

John Von Neuman. I knew him well. About twice a year Johnny would phone me and we would compare travel plans. We would select a time and place where just the two of us could be together for a half-day or day, and we sometimes shared the same hotel room. His untimely death was a tremendous loss not only to the nation and the world, but also to me personally. I could tell much about him, but now just one item: When he returned from his first trip to the far north (Thule, Greenland) he commented that his biggest problem was wearing six inches of clothing, particularly from the waist down.

* * *

Theodore Von Karman. I did not know him well, but I knew a lot about him. He established the Air Force Scientific Advisory board at the request of Hap Arnold, and was its first chairman. He published a treatise predicting the future contributions of science to military capabilities, which was, and still is, a classic; on 19 December 1945 the summary volume "Toward New Horizons" was given to Hap Arnold, along with some 20 supplementary volumes. He established AGARD (Advisory Group for Aeronautical Research and Development) in NATO. He was single and quite a ladies' man. The first time he met Guy Stever's wife, he exclaimed, "Wouldn't you know that a Guy would be married to a Doll?"

* * *

H. Guyford Stever. I've known Guy well since early in World War II. Among his activities, he has been with the Office of Scientific Research and Development (OSRD), chief scientist of the Air Force, dean of the Graduate School at MIT, president of the Carnegie-Mellon Institute in Pittsburgh, and director of the National Science Foundation.

While he was president of Carnegie-Mellon, and during the height of the college students' uprisings, one day while having lunch with him I asked about student problems at Carnegie-Mellon. He replied that they had had very few. Then he told this story: April Fool joking had become tradition in his family. One April Fool's Day, shortly after arriving in his office, he phoned his wife and said, "Well, it has happened. I'm a prisoner in my office." Her first words were, "They haven't hurt our picture have they?"

* * *

Physician. Late one night a plumbing difficulty developed in his home. He had a casual acquaintance with a plumber who lived down the street. He phoned him and the plumber arrived shortly. After the problem was eliminated, the physician said, "Just because we are neighbors, I don't want any special favors. Charge me your usual fee." So the plumber said, "Fifty dollars." Taken somewhat aback the physician said, "I get only twenty-five dollars for a house call." The reply, "That's all I got when I was a physician."

* * *

Lasers. For ten months in 1970 I was director of the USAF Mission Analysis on CONUS Air Defense at Wright-Patterson AFB, Dayton, Ohio. A mission analysis has two main products: (1) operational concepts (these embrace weapon systems, men, communications, support, procedures, and so on; I had difficulty with some Wright-Patterson types convincing them that a weapon system was not an operational concept), and (2) an assessment of the technological "state of the art" to suggest research and development required to attain the identified and defined operational concepts.

Some previous mission analyses had fallen flat on their faces because they gave insufficient attention to product #2, technological state of the art. I had personally heard members of the Air Force Laboratories and the Scientific Advisory Board tear those efforts apart, and not very gently or politely. I was determined not to make that same mistake. Many of the Air Force Laboratories were at Wright-Patterson. I convinced, persuaded, coerced the labs there and elsewhere to write my technical annexes for me, and I was able to arrange for some of the draft annexes to receive informal reviews by selected members of the Air Force Scientific Advisory Board. But throughout my career I would not sign anything that I personally did not understand, or was able to support. Hence I received many detailed briefings from members of the various labs. Once I had a full day scheduled for briefings by four representatives of a lab. The last briefing was to be on lasers. It turned out that the previous three briefers each spent considerable time on lasers, though it was not the main subject for any of them. Finally, the gent to brief on lasers got to the floor. He said, "I feel like Elizabeth Taylor's next husband. I know what I'm supposed to do, but I wonder if I can make it interesting."

* * *

Nightgown. The woman in a drygoods store was selecting material to make a nightgown for herself. She found some she liked and told the clerk she wanted 40 yards. The cleark asked, "You said 40 yards?" "Yes, 40 yards." "You mean 40 yards?" "Yes, I mean 40 yards. I'm married to a scientist and he finds it more interesting to look for it than to find it."

* * *

Decision Theory. An acquaintance of mine, a nationally known economist at Columbia University, told me this story: A colleague at Columbia was an expert in mathematical logic — Decision Theory; if a Nobel Prize were awarded in that discipline, he would receive it. He got an offer from another university, and he had difficulty deciding whether to accept or reject. One day he asked his economist friend to take a walk with him on the campus for him to think out loud. He expounded on the pros and cons. Finally his economist friend asked why he did not use some Decision Theory. The reply, "Charlie, be serious. This is important."

Chapter 17

Diplomatic Discernments

Although I have been on the periphery of various diplomatic deliberations, the only time I was officially involved in diplomacy was in Spain. At that time I was with Hq SAC in Omaha. I was selected as an official member of a three-man team to negotiate for the USAF with the Spanish the renewal of the use of SAC bases in Spain. The other two members were major generals from Hq USAF, the Pentagon. (In 1974 it required Henry Kissinger, then Secretary of State.) We were in session seven straight days, including the weekend. Each morning we met with the American Ambassador to report important decisions of the previous day, to speak of the agenda for that day, and to receive any pertinent guidance.

All the negotiations, and discussion leading thereto, were translated. The Spanish pretended they knew no English, and it was a fact that I knew no Spanish. Each day was a lengthy session. We completed mutual accord, and we and the Spanish signed the agreement, and when the official business was completed, we were guests of the Spanish Air Force Chief of Staff for dinner at Horcher's, the famous German restaurant in Madrid.

The three members of the American delegation each had an advisor, so when we were ushered into a private dining room, there were twelve of us, six plus six. By actual count there were two waiters per diner, and we had a wonderful seven-course dinner with three different kinds of wine. I was much amused, having heard the Spanish Air Force Chief of Staff order the seven-course dinner for all of us, except for him. He had filet mignon, baked potato, and green salad. You will tend not to believe this, but it is true: he also had bottled water.

By mid-meal most of our hosts could speak excellent English. One of our advisors, commenting on the food, said it was not hot; he had expected it to be Mexican-like. A general from the Pyrenees, the northernmost part of Spain, explained that hot Spanish food existed only in the south along the Mediterranean, and that both bullfighting and Flamenco dancing were imported to Madrid and the north from southernmost Spain. That started a discussion of the different cultures in the northern and southern parts of a given country. We agreed that differences existed in the United States, in Spain, in Italy, in Germany, even in England, in France, and others. At this point the Spanish general from the north said, "I remind you that southern France is north of northern Spain."

The senior major general and I shared a deluxe suite at the Castillona Hilton Hotel in Madrid, a tremendously large, elegantly furnished set of rooms. Fortunately, we received military rates. After each daily session my suite-mate and I had the rest of our party in for a drink or so, before we went to dinner, to decide which part of the city was going to receive our attention that night. My suite-mate and I had agreed to share the expenses for these "happy hours." I got my pesatos from the Embassy at a much better rate than one could get from a bank. The exchange rate at a hotel was considerably worse than from a bank. My suite-mate paid for his half with an American Express card. Half the expenses cost him over $100 more than it cost me, because of the difference in exchange rates. This did not bother him at all, as he had had the foresight to marry a very wealthy woman.

* * *

One of the bonuses of my positions at SAC was the opportunity to take part in meetings with many distinguished visitors, both American and foreign. I was always one of four or five staff members to join General LeMay in such sessions. Typically, they would arrive just before dinner. We would have dinner with them, and drinks afterwards, getting acquainted. The next morning there were sometimes briefings at breakfast, then sessions, lunch, sessions, dinner, and sometimes sessions after dinner.

Among the people whom I met at such sessions were Paul Henri Spaak, United Nations General Assembly President, the Shah of Iran, the King of Morocco, Viscount General Montgomery, the British Shadow Cabinet when the Labor Party was in power, the U.S. Joint Chiefs of Staff, Italian, French, German dignitaries. Also many Americans, for example, Wilson, Gates, and McElroy when each was Secretary of Defense.

One visit was from the Department of State, when the Russians were holding a crew of a SAC RB-47 as spies. John Foster Dulles and Herbert Hoover, Jr., were there. It was the first time that the Secretary of State and the Under Secretary were out of Washington at the same time. Henry Cabot Lodge, our Ambassador to the UN, and Mr. McArthur (no relation to the general but a wheel in the State Department), and others were there.

It happened that I sat beside Lodge at both lunch and dinner. I was delighted when he never said, "the Russians" or "the Soviets." Instead he sad, "those bastards" or "those SOBs." He said that an important part of the Soviet operations was to lie. Some, with documentation, would know they were lying, but most of the people in the world would not.

Lodge said that it was extremely important to have an interpreter who could not only translate Russian into English, but also had a brain, to point out important points and subtleties. He was fortunate to have one such. Early in Malik's first tour in the UN in speaking of the RB-47 crew he referred to "the Americans in uniform." Lodge's lad caught this immediately. When Malik finished, Lodge pointed out that he had said, "Americans in uniform." Therefore they could not be spies. Though Malik knew it had been recorded, he denied he said it. That night Lodge's lad taught him the Russian words for "Americans in uniform," and next day Lodge said them in Russian at the UN. Within a few days the RB-47 crew was released. Note the Russian reaction to determination. Remember Lebanon, Quemoy, Cuba.

Lodge said that though he enjoys a drink now and then, cocktail parties were a hazard to his profession. But he usually had no choice but to go. At one he was talking in the corner with two quite important Russians. One asked if he knew any Russian. He repeated his bit about "the Americans in uniform." They laughed, guffawed, beat their knees, and slapped him on the back.

* * *

On my first visit to Vienna after World War II I was met by the USAF air attaché and his wife. That night the senior military attaché, an Army officer, was having the diplomatic party of the year. He lived in a magnificant old mansion, almost castlelike, and important people from all the embassies and legations in Vienna were there. The Army host introduced me to many guests. One was the senior Russian there. When I was introduced the Russian asked, "What kind of important general are you that they introduce you as Mister?" I did not reply. Then he turned to the gent from Luxembourg and said, "My! Your hair is getting so gray."

The reply, "My! You do notice everything, don't you?"

The Russian turned to me and asked, "What are you doing in Vienna?" I made a reply that I thought was great: I said, "I'm visiting." After a look at me with some emotion, I could not tell if it were admiration or disgust, he pursued the point no further.

* * *

Though he retired ten or more years ago, Dr. R. P. Zimmerman for years was in charge of the Canadian military activities with respect to research and development and systems analysis. Each year for quite a few years he invited some of us to visit Ottawa. There was no official business action purpose, but we were providing his scientists and analysts the opportunity to make presentations on the work they had accomplished to a critical professional audience. Although a civilian, he was a member of the Canadian Joint Chiefs of Staff, and on his visits to Washington the Canadian Embassy often held a reception and party in his honor. Modie and I were always invited. It was quite interesting what some guests said to us, who did not know either Dr. Zimmerman.

* * *

In 1967 at a party at the Italian Embassy Modie and I met a Polish officer whose wife had just arrived, though he had been here for 10 months. She was a doll, but if she knew any English she would not demonstrate it. He told us that Polish officers had to be fluent in five languages — Polish, English, Russian, French, and German. When it was time to move along, I asked him to tell his wife that she had beautiful eyes, and beautiful other things. He laughed. What he told her, I don't know, but she smiled demurely.

* * *

In February of 1962 I was visiting Portugal. I wondered what kind of reception I would get, as an American citizen, because at the two most recent opportunities in the United Nations, the United States had voted against Portuguese interests. Late in the afternoon of the twentieth I was with our Ambassador to Portugal in his office in Lisbon, when he received a call informing him that Lt. Col. John H. Glenn, Jr., was successfully in orbit.

Eating dinner at the dining room in my hotel I heard yelling outside. A waiter brought me a newspaper extra that was being sold on the street. It didn't do me a hell of a lot of good as it was written in Portuguese, but it was clear that the natives were excited, and highly in U.S. favor.

After dinner the bell captain was listening to a radio.

He spoke a smattering of English. He asked if I understood Portuguese. No. French? Yes. So a Portuguese bell captain proceeded excitedly to tell me in French the highlights of Glenn's flight.

* * *

In 1974 we were at a party at the Japanese Embassy honoring the Chief of the Japanese Air Force. A major general in his party, Operations, sought me out because he had been told to do so in Japan before leaving. He gave me personal regards from some individuals in Japan. The Japanese serve excellent food at their embassy parties.

Dr. Lauriston S. Taylor, Chief, OA, Ninth Air Force and IX Bomber Command during World War II. Retirement banquet from the National Bureau of Standards, December 1964.

With General Thomas White, Chief of Staff, USAF, and Mrs. White, early 1970s.

With Maj. Gen. Glenn Kent, Hq USAF, 1970.

Receiving an award from General Chapman, Air Force Systems Command, or contribution as Director of Mission Analysis for CONUS Air Defense. I was on loan for 10 months in 1971 to the Air Force from Autonetics, a division of North American Aviation.

With General Fred Smith, Vice Chief of Staff, USAF, 1970s.

Chapter 18

Remembrance and Reverence

Anyone who has any degree of success in my profession (variously called Operations Analysis, Operational Research, or Systems Analysis) owes that success to many individuals. One category is those who have worked for him. I was fortunate to have excellent professionals around me. Some who made significant and outstanding contributions to me, and therefore to the Air Force, are no longer alive. I recruited all but three of them into Operations Analysis. Those three had been in Army Air Corps Operations Analysis, and I selected them to continue with me when I returned from Europe after V-E Day.

Only two were slightly older than I when they died. Most of the rest were considerably younger. I owe much to each of them. They are listed here in an approximate order of their demise.

* * *

Harold Balmer: I knew him first at the Radiation Laboratory at MIT in early World War II. He joined me overseas and carried out many fine jobs there. Two outstanding ones were directing the mutual interference tests among U.S. electronic equipments to be used in the invasion, and assisting in developing procedures for, and in training of, U.S. Pathfinder crews. Both tasks were done before D-Day. When SAC was originated after the war, he joined my Operations Analysis Office there.

* * *

Harry Chittim was an educator from New England. When I returned after two years in Europe during the war, to become Chief of the Operations Analysis Office of Continental Air Forces, I inherited all the analysts in the four numbered Air Forces under CAF. I decided to keep only two of them. After the war in the Pacific ended, many wanted to return to their pursuits before they became involved with the Army Air Corps. One I kept was Stan Lawwill, who became my deputy and was with me at SAC for 13 years. He recently retired as President of Analytic Services, Inc.

After perusing background and contributions during World War II I decided I wanted Harry Chittim. He was collecting his salary and doing various things at various installations. I could not contact him. I phoned, sent letters and TWXs. No reply. Finally

Orientation in Systems Acquisition for General/Flag Officers, Defense Systems Management School, Fort Belvoir, Virginia, March 1974.

With Dr. Marlin S. Reichley, Senior Educational Advisor, Industrial College of the Armed Forces, Washington, January 1974.

about the middle of a week I sent a TWX to his attention at three different places. It said that as of Monday next, his services would be terminated. I heard from him immediately.

He did a fine job in steering, and developing SAC training practices and procedures.

Harry could play the piano extremely well, but did so only after two or three drinks. One night at the officers' club at Andrews AFB he started playing late in the evening, after the dance band had left. Almost all activity stopped, and people crowded around the piano — it was his first performance there. A colonel said to me that I ought to hire him. I replied that he already worked for me.

When Rod Smith succeeded me as Chief of Operations Analysis he selected Harry as his deputy.

* * *

Dayle Rippe: I made a remarkably successful recruiting trip, concentrating on Big Ten universities. I wrote to heads of Physics and Mathematics departments and graduate school deans, describing my needs and requesting suggestions for candidates. I hired Dayle from the Univerity of Michigan. He was a bright, eager, and capable young gent.

Once on a trip to California he, Rod Smith, Tex Alexander, and I went to the Colony Club in Compton in the environs of Los Angeles. For that era, the girlie shows were great. As we were leaving, Dayle asked, "Do the police know what's going on there?"

After a few years Dayle joined IBM in New York City.

* * *

Tex Alexander: The dean of the graduate school at Purdue recommended Tex highly. After several years with Hq SAC in Omaha he served in Operations Analysis at Hq USAFE, Wiesbaden, Germany. After his tour there he joined me at Hq USAF, where I had gone from SAC. Later a vacancy developed for Chief, Operations Analysis, at Hq Pacific Air Forces at Hickam AFB at Honolulu, and I selected Tex for the post. After a while he suddenly died of a heart attack.

* * *

Gene Jackson: For a vacancy for an electronics expert, Gene, his boss at Wright-Pat AFB, and several others were candidates. Upon joining me he said he did not understand why I chose him rather than his boss. After some years of outstanding contributions to SAC I said, "Now you understand."

He replied, "No."

* * *

Paul Hower: Paul was a physics student of mine at Whitman College. Upon graduation, during World War II, he went to the MIT Radiation Laboratory. When I assumed my position of Chief of Operations Analysis at SAC, I was looking for someone with demonstrated expertise for certain responsibilities. I was unsuccessful in getting such a person, so I decided that perhaps even better would be a young capable person with potential. A bell rang, "Paul Hower." I went to the Brooklyn Polytechnic Institute, where he had gone after the war, and hired him. He was outstanding as a team chief and Rod Smith selected him to succeed him as Chief, Operations Analysis. Again, Paul did an outstanding job, and he became my successor as Chief, Operations Analysis, USAF, in 1962. There, again, he was outstanding.

While going to an Air Force Association convention in San Francisco, Paul had a heart attack and died a few hours later.

* * *

Rod Smith: Scott Rethorst from MIT strongly recommended Rod, Assistant Professor of Aeronautical Engineering at MIT. I hired Rod to come to SAC for a year; he stayed 11 or 12 years. We learned, after he arrived at SAC, that both he and I had been scheduled to go to Sao Paulo, Brazil, to help them get their equivalent of Wright-Pat established. The only requirement for the renewal of the contract, except for not falling flat on one's face, was to be able to write, read, and speak Portuguese after a year. It was a challenge, and the salary was excellent. For different reasons we both decided not to go.

Rod was an outstanding professional, human being, and gentleman. He was highly respected by industry, the military, and academia. Dr. Robert Seamans, director (in 1976) of the Energy Research and Development Agency and former Secretary of the Air Force, now dean of the School of Engineering, MIT, told me that Rod was one of the most outstanding young men he had known.

I made him a division chief and then — when I became Chief Scientist — Chief, Operations Analysis. We knew it was only a matter of time until Rod would have to take care of considerable family land holdings from Los Angeles to San Francisco, and eventually he did resign from the Air Force to go to California. When I went to Douglas Aircraft in Santa Monica in 1962 Rod became a consultant to me and spent one day a week with me, or longer when necessary. He came once just after Christmas and worked for five straight days. This included New Year's Eve, but Rod and I were in the office at six in the morning on New Year's Day.

When I resigned from Douglas and went to Autonetics, North American Aviation, Anaheim, he joined me there as a consultant for one day a week. For quite a few years I spent more time with Rod than did anyone else except his family.

We gave a large party for friends at our home at Cameo Shores, but I got a message that Rod could not come. Rod had not known that anything was wrong with him, but after his annual physical the doctor told him he had terminal leukemia and about two weeks to live. He died on New Year's Eve and I was pall-bearer at his funeral on January third.

* * *

I hired either people that I knew, or that people whom I respected knew. There were several lineages in my office. One example was that through Scott Rethorst from MIT, I got Rod. When Rod left MIT, Mart Schwartz was put into his position as an assistant professor of Aeronautical Engineering. A year later I hired Mart. Then John McCarthy was put in that position. A year later I hired John. No prejudice from MIT. I've known all but one of the MIT chancellors well, starting with Karl Compton, Jules Stratton, Jim Killian, Howard Johnson, and Jerry Wiesner. I did not know Johnson. Karl Compton knew me slightly but Stratton, Killian, and Wiesner called me "Zim."

* *

Campbell H. Watson: Cam had one of the best minds of anyone I have known. He was independently wealthy. He was my deputy at Hq SAC just after World War II. When we moved to Omaha, he chose not to go, but when I went to Douglas in 1962 he rejoined me as consultant and continued with me at Autonetics. He went into a hospital in Los Angeles. For some days we had not gone to see him, but Modie said we must see him that night. We did. He died about noon the next day.

* * *

George Nicholson was another person I inherited. He was chairman of the Department of Statistics at the University of North Carolina, and director of the Operations Analysis Standby Unit there. In that capacity, and as consultant to me, he made many contributions. He was technical director of the G-91 interceptor tests in Italy. This was an Italian aircraft in which the West Germans were interested. He knew his way on the Hill and was a confidant of Senator Henry Jackson of Washington state. Again one wonders why one so young, with much to offer the nation and the Air Force, is no longer with us.

* * *

Dan Leivick: When I was in the Pentagon from 1959-1962 I made Dan one of my team chiefs and secured a supergrade for him. When I returned to the Pentagon in 1970 he was still in that position. He retired about a year later.

His father was a famous Israeli poet and playwright, though he lived in this country most of his adult life. One of his claims to fame was writing plays in Yiddish, which played on Broadway. He may have been the only one to accomplish that. In Israel the equivalent of our Kennedy Center was named in his honor.

Dan and Ida knew Golda Meier and were her house guests more than once. I used to embarrass Dan on occasion when, in his presence, without saying anything about Golda, I would ask someone how many men he knew who had kissed a Prime Minister.

* * *

Wendall A. Dwyer died of a heart attack in Seattle in August 1977. Wen had been a professor of Mathematics and then assistant to the president at Creighton University, Omaha. When SAC moved to Omaha he had gone to Washington University, St. Louis. I hired him to join me at SAC.

Wen was an outstanding professional. He had a great rapport with the military and was very considerate and helpful to those who worked for him. For years in early SAC he made outstanding contributions to developing training procedures and policy, and he assisted in a major way in developing tests and training exercises. He was in charge of analyzing the results for SAC.

* * *

Frederick F. Fennema: During World War II Fred Fennema was an operations analyst with the Twentieth Air Force in the Marianas. Immediately after the war he was with the Alaskan Air Command as an operations analyst for several years. He joined SAC after the headquarters moved to Omaha. He suggested and fostered the B-37 low level penetration and toss bombing; he was a key member of the B-36 operational engineering program; and he was heavily involved in Project GENETRIX.

After leaving Operations Analysis, Fred completed his Ph.D. and became a professor at Cal Poly at San Luis Obispo. He died in late 1982 in Texas.

Chapter 19

Reflections About General LeMay

I retired from the Air Force, as a civilian, in 1975, after 25 years with the Air Force, during over half of which I reported directly to General LeMay. Through the years there have been many stories about General LeMay. Most of them are untrue. The following stories are true.

When Frank Pace was Secretary of the Army and Dan Kimball Secretary of the Navy, they visited Hq Strategic Air Command together. They came from Washington in the same aircraft and played gin rummy on the way out. Upon landing at Offutt AFB, Pace said to General LeMay, "I won a carrier from Kimball on the way out." Without a second's hesitation General LeMay asked, "The cost or the value?"

It was during these meetings that at one point General LeMay said to Secretary Kimball, "Now, dammit, Mister Secretary, Sir."

A major general on the staff of General LeMay was reporting on the results of his visit to the Pentagon. Quite satisfied with the accomplishments on his mission, the general summarized, "We screwed them on that one."

General LeMay quietly replied, "Just make certain that in a few months, you're not pregnant."

One Saturday morning in Omaha I drove through a blizzard from my home to Hq SAC, arriving at my office at 0820. The white flag on my squawk box was showing under the Commander's position. This meant he had tried to call me. I immediately went to his office and asked, "Did you want to see me, Sir?" He replied, "Yes, I've been trying to get you all morning." Yet on several occasions he introduced me to some rather important individuals as "the brains of SAC."

* * *

Here are some comments from others who worked closely with General LeMay (the officers are now retired):

Thomas Chappelle

General LeMay was a very fair man who gave a person ample opportunity to demonstrate competence. On the other hand, I witnessed examples of his low tolerance for incompetence and failure to follow directions. One example is especially vivid in my memory. The occasion was a briefing to him by a delegation from Wright-Patterson AFB in the command conference room. The briefing was a follow-on to an earlier one at the end of which General LeMay had asked the WPAFB team to look further into some subjects and then report back. The briefing officer started in, but he had covered only two or three charts when General LeMay uncharacteristically interrupted to ask a question related to the additional information he had requested earlier. The briefing officer waffled, not actually answering General LeMay's question. General LeMay quietly asked him the same question in slightly different words. The senior officer in the WPAFB contingent took over, gave an equally waffled, unsatisfactory non-answer, and told the briefing officer to continue. Before he could do so, General LeMay looked the senior WPAFB officer in the eye and said something like, "Colonel, I am going to ask a direct question and I want a direct answer. Did you or did you not do the things I asked you to do during your last visit?"

The senior WPAFB officer hesitated a second and then said, "No, General. We did not."

General LeMay stood up and said, "Gents, this meeting is over," and left the room.

Another interesting characteristic was General LeMay's style of receiving briefings. In my experience General LeMay almost invariably sat quietly throughout a briefing, giving no indication of whether he thought it good or worthless, whether he agreed or disagreed with its contents, or whether he thought the briefer was stupid or a genius, unless you could interpret the smoke signals from his ever-present cigar. Some briefers found his style very disconcerting, even though predictable.

General Power, on the other hand, was completely unpredictable. He might interrupt on the first chart and keep asking questions and commenting, with the result that the briefer would never get back to his prepared presentation except occasionally to thumb to a chart to illustrate a point. However, in the next briefing session, General Power might follow General LeMay's impassive procedure. With General Power, it was a very sporting course!

General LeMay was no golden-tongued orator, but — especially in a question and answer exchange — he frequently excelled. I well remember an Air Force Scientific Advisory Board meeting that the Operations Analysis Office had arranged for and managed. General LeMay was in top form that day and enjoyed a lengthy discussion with the Board, whose members were obviously impressed extremely favorably by General LeMay's insight.

One last subject: General LeMay's proverbial taciturnity. Every so often, he broke out of that mold, too. An occasion I especially remember was an afternoon command briefing to some VIPs. I can't remember who the VIPs were, but they were on the level of Congressmen or a Cabinet member and his party, or it could have been the chairman of the Atomic Energy Commission and his party, or someone similar. In any event, Maj. Gen. J. B. Montgomery was the briefing officer and General LeMay was sitting down below with the guests. The visitors had a departure time that set the time limit for the briefing session. Monty did a superb job in those briefings. That day, however, General LeMay kept interrupting and enlarging at length on points Monty was making. About midpoint in the briefing, after an especially time-consuming interruption with the clock ticking away, Monty hesitated before resuming his presentation, looked at General LeMay, and said something like, "Boss, if *we* don't limit our elaboration of all these interesting subjects, we're just not going to get through this briefing for these gentlemen."

Lt. Gen. Richard M. Montgomery

My time in SAC seems so long ago, but I remember much of it as though it happened just a few days ago. I recall very vividly the great contribution that the operations analysts made to our SAC mission.

I reported for duty at Hq SAC in March 1951 as deputy to the Chief of Staff, Maj. Gen. Aug Kissner. I had served with him previously but had not seen him for many years. It was a wonderful opportunity to work at this high level with such skilled officers and with an old friend. But it was a terrifying experience to go in and face General LeMay the first few times. He fixed his eyes on you and sat quietly waiting for any gems of wisdom to drop from your lips. I didn't have much to say the first few times. I was really there to listen and learn. He is a great man, and in short order I learned to interpret the gestures he made — the tilt of his cigar, the raising or lowering of his eyebrows; I could tell his mood by the way he sat in his chair. After a while I learned the signals and could guide the staff in or out — when it was best to go in, or when not to disturb him whenever he was wrestling with a serious Air Force problem.

Many times he had to make a trip to Washington to rescue the B-52 bomber aircraft. It seemed like someone was always cutting the B-52 out of the program. General LeMay knew that a great defense was fine to have, but was secondary to the offensive punch needed to sway the enemy. He became deeply concerned from time to time over the lack of progress in the B-52 program. We were very fortunate that we had such a great leader as LeMay, or we would not have the B-52 today. The B-52s have kept the peace for 25-30 years. I don't know yet if we will get the B-1 bomber. I pray we do. It's long past time for replacement of the B-52.

I had been in SAC Hq only six months when Aug Kissner went on temporary duty to the Azores to negotiate with the Portugese for a base in the Azores. That lasted for months, and Aug did an outstanding job. Then he moved on to Spain to set up negotiations for the USAF to build SAC bases in Spain. Incidentally, those bases are the finest ones in Europe, even today. Aug Kissner did a great job for SAC in Spain, when he had a 500-mile pipeline built from the Mediterranean to northern Spain. I think that period really put the Spanish on their feet.

I was in deep water to act as Chief of Staff, with Aug away, working with the other top staff officers, all of whom were senior to me. Also with Tom Power, who was a major general at that time, and deputy to General LeMay. When I reported General LeMay had three stars, but shortly thereafter was promoted to his fourth star. It is not generally realized that LeMay and Ulysses S. Grant are the only two general officers who received their fourth star prior to the age of 45.

Tom Power worked very closely with LeMay; they made a great team, and it was difficult for me to keep up with those two dynamic people. But gradually I learned my business — when to go in and talk with LeMay, and when not to, and also how to handle the staff. There were times when he needed to be alone thinking about the future, and at those times I would tend to ask the staff to stay away, unless it was really important.

Often he would ease the tension for 30 minutes on the firing range. He was an excellent marksman. Butch Griswold, who was his deputy after Tom Power, would often go with him. I remember that when the Provost Marshal of the Air Force, a major general, came to visit SAC General LeMay took him to the range. The visitor hardly knew which end of the gun the bullets came out, and he got a good dressing down. How could he be a good Provost Marshal if he didn't know about his arms? The Air Police came under his cognizance, as did the security of Air Force bases, including SAC bases. He seemed to be an armchair Provost Marshal rather than an active one. LeMay made him familiar with the survival programs SAC was conducting in the mountains, while no one else in the Air Force was paying much attention to them. We had outstanding programs for crews in case they were forced down in enemy territory. Stead AFB, Nevada, and later Colorado Springs were the sites for this training.

A reserve general officer who was a classmate of mine at flying school came in and asked for an opportunity to brief General LeMay on the Reserve Program. I set it up. Before we went in he said, "I have never met General LeMay and I'm very nervous." I explained to him that LeMay would listen carefully. My friend could count on the time allocated, uninterrupted. When LeMay focused those eyes on him and just kept looking, the briefer got so excited and upset, he could hardly talk. Finally he said, "General LeMay, I have never met you before. Just being here trying to brief you is so upsetting I can hardly think." LeMay did something that was very characteristic of the man, but something few would expect. He said, "Keep on; you are doing all right." He sat back and shifted his cigar, and the briefing went fine from there on.

I recall numerous briefings that the SAC staff and later the Air Staff had labored over for days. LeMay would sit quietly and listen. At the end he would point out the weakness or deficiency of the study, and the staff would sit stunned. "Why hadn't we thought of that?" He had the innate ability to analyze and ferret out omissions or shortcomings on first observation.

General LeMay is a man of great compassion, although he does not have that reputation. He is a warm-hearted, kind, and very understanding person, very loyal to those who were loyal to him. I once heard him telling a visiting civilian, "I have men on my staff that I don't like worth a damn, but I like their work." That was another measure of this great man: he could rise above personalities, above his personal like or dislike of an individual, to get outstanding performance and professional judgment from the best men he could find. In my over 30 years of service in the Air Force I have encountered very few people who could overcome their prejudices in this way. We must admit we all have some prejudices.

Another example of his compassion: I had a serious problem with my young son who was rather severely retarded. He had to be taken to a special school in Pennsylvania. I had proposed flying him in a small airplane which General LeMay and three of four of us had taken apart and rebuilt. It would have been a seven-hour elapsed time from Omaha to Philadelphia. In that light plane over the mountains my son would have been quite upset, with the length of time in the airplane, the strange environment, plus the several stops for fuel. Tommie might not have been able to handle the changing situations, and could easily have got into a very serious state, as retarded as he was. I was quite concerned about getting him to Philadelphia.

When General LeMay learned what I had planned to do, he authorized for my use a C-97 with staff officers going to Washington; we let them off, and then took Tommie to Philadelphia. I will never forget his deep understanding, kindness, and warm-heartedness.

It is clear how I feel about General LeMay, one of the greatest men our country has ever produced. I worked very closely with him for about eight years. One of his greatest concerns was the welfare of his troops. He got Senator Wherry of Nebraska to sponsor legislation which later became Wherry Housing, the first civilian-sponsored housing authorized on a military base. A few years later, when it proved that that money was not enough, Capehart Housing authorized more dollars for each house and turned out an even better product. General LeMay had contacted Senator Capehart of Indiana.

He established hobby shops, where one could fix one's own car. If you didn't know too much about it, the equipment was there and a qualified sergeant would show you how to use it. Many of our aircraft mechanics were excellent; they could do a better job repairing a car than most of the mechanics downtown. All they needed was equipment and a place to work. These facilities were established at all SAC bases.

General LeMay led the way by producing a little sports car. He published a manual on how to build it: take this kind of a part from that car, and that kind of a part from this car, and a body. He thought that many of his airmen could put their money and effort into a car, instead of spending it weekends drinking beer. It would make them better mechanics, and when the year was over they would have a car they could register, and it would operate very well.

One time he had his Cadillac at the hobby shop at Offutt AFB. He was in a pair of GI coveralls, no insignia, and was on his back underneath the car. A new sergeant came in. Seeing no stripes on LeMay's sleeve, he asked "What are you trying to do?" General LeMay said, "I'm trying to get this bolt out."

The sergeant observed, "No wonder you have a clean sleeve the way you handle that wrench." The other airmen around nearly had apoplexy. But General LeMay thought that was one of the funniest things that had happened recently. He roared with laughter, but never revealed his identity.

He has a great sense of humor, something not many people would suspect. I enjoyed taking a good joke to him once in a while, just to hear him laugh.

Another thing that was impressive was his willingness to try something new. He became very interested in new equipment. He and then Major General Griswold, his deputy, had been ham radio operators for some time, and they became quite interested in single-sideband radio possibilities. They equipped several of our SAC aircraft with single-sideband radios. Butch flew one up over the North Pole. He could talk to almost anyone in the world. In short order, the entire SAC force was equipped with these unusual radios. We wanted to get combat reports back from our crews. With this equipment, when it came time to break radio silence over the target, the crew could report combat strike results directly to Hq SAC.

I had command of the SAC Third Air Division on Guam for a while in the late 1950s. We had single-sideband. A big sign over the door in the radio room aid, "Before you start talking remember anyone within 10,000 miles can hear you."

I was well out over the Pacific flying a B-47, and in rapid succession talked to Alaska, Hawaii, and Guam. It was as clear as a telephone. While I was on Guam the phone rang one evening. LeMay's familiar voice was unmistakable. He said, "I'm in my room in a Hong Kong hotel. I have this single-sideband package which I can carry. The antenna is a wire which I hung out of the window. How are you receiving me?"

I was glad to report, "Five-square, five-square, Sir." He was real pleased. LeMay tried much of SAC's equipment himself, even flew every type aircraft in the force.

One day a young captain was in before LeMay to explain why he landed his B-47 on a 4,000-foot strip instead of the Atlanta Airport. The strip was in line with the long runway at Atlanta, but about five miles short. LeMay said, "Young man, didn't that runway look short?" The captain said, "Sir, all runways look short when you are landing a B-47." I thought General LeMay was going to fall out of his chair. He roared with laughter. That ended the conversation. (The lad had landed the B-47 safely on a 4,000-foot runway when no one believed one could land a B-47 in that short distance; the wings had to be removed and the fuselage towed five miles for takeoff.)

General LeMay built and commanded the greatest force for peace in the world. There has never been such an effective force since then. He always wanted Operations Analysis inputs in all the tough decisions, especially in war plans.

Over a period we noted that more than 50 percent of General LeMay's time was spent trying to better conditions for his troops. He thought better housing should be available for families, and better barracks for single personnel. He got the Sports Car Club of America to conduct races at various SAC bases. For an admission fee the local people could see first class cars being raced, and we raised between $60,000 and $100,000 at each race. That money was used to buy furniture, drapes, covers for the beds — everything to make a more homelike atmosphere in the barracks.

[Author's note: A member of the United States House of Representatives got the races canceled. He claimed General LeMay was causing airmen to work without pay on their free time. Hell, by far most of them enjoyed it. I know, because I was there. Thus a source of income for the airmen's betterment was canceled.]

General LeMay also had SAC barracks designed with small rooms — only two, or sometimes three, men to a room — with a wash bowl, where the men could wash and shave, and a medicine cabinet and mirror. Between each two rooms was a toilet and a shower. This is another example of the things he did to try to keep the SAC force more comfortable.

[Author's note: General LeMay was widely criticized by some for coddling his airmen and fostering homosexuality with only two or three men to a room. Many dedicated airmen and oficers deserve much credit for staying with the Air Force when continually faced with such nonsense from naive and lame-brained (reference the military) persons in positions of authority. Why would a self-respecting member of the Air Force suffer these kinds of things? Dedication to our nation and what it provides for us.]

General LeMay was a truly great man. I watched him assaulted by Secretary of Defense McNamara and the do-gooders, people who professed to know so much, though they had never been involved in the defense business before. Some made every effort to cut him to ribbons. However, he was above it all and was highly regarded by the Congress. Though held up to ridicule he faced it very well. He was pushed around by McNamara's band, small people who seemed to take great pleasure in trying to demean and belittle this great man. He rose above them all.

He ended his career as Chief of Staff, United States Air Force. I served as Assistant Vice Chief of Staff when he was in the Vice Chief's position, and for a year after he became Chief. It was the greatest experience in my life to watch up close this great man at work.

Col. Lawrence J. Israel

In my four and a half years on the staff of General LeMay I learned many things from him. The most important was total honesty. I believe that if someone put a gun at his head and told him to lie, he would have to shoot, because I firmly believe General LeMay would not have known how to tell a lie.

I learned also to never fabricate, or partially fabricate, an answer to a question. I witnessed the departure of a major general from Hq SAC one afternoon because he attempted that. Some others had hasty departures. When he asked me something I did not know, I would say, "Sir, I do not know. But I will get the answer in ten minutes (two hours, two days, or whatever)." In each instance he said, "Very good. Go do it." His acute sense of judgment and discrimination permitted him to know when he was being subjected to a fabrication.

His search for knowledge extended to the lowest levels. With experience in industry and the Air Force I have never experienced such complete honesty elsewhere. This includes the academic world, as I taught mathematics at a college for three years upon retirement. His willingness to listen when he thought one had something to say was outstanding and, perhaps, unique. I was never denied permission to enter his office. He listened to me, and to my people, and was an open vessel for the reception of knowledge. He frequently asked perceptive and pertinent questions, but he listened. General LeMay was the greatest man I have known.

I must also acknowledge a great debt of gratitude to General Power. I like to think he was a personal friend of mine. He was of inestimable help in taking me over the rough spots in my dealings with General LeMay, and of course there were rough spots. General Power steered me in directions I should go, which I might not have recognized without his guidance.

General Jack J. Catton

As a young lieutenant colonel on General LeMay's Twentieth Air Force staff responsible for mission planning, tactical doctrine, and writing operations orders, I was tremendously impressed with his complete knowledge of the capabilities of his aircraft, systems, and crews. His extensive experience and personal participation in developing strategy, organizing forces, and applying them gave him extraordinary insight into the performance of his extremely important responsibilities. Of particular note: his personal frequent air leadership of the most important missions over the toughest targets. He was forthright, intolerant of mistakes, decisive — and fortunately, he never made a bad decision in the use of his forces in combat.

After successfully carrying out the Berlin Airlift, General LeMay returned to the United States to become CINCSAC. In a few short months he led intensive planning and program efforts which provided the blueprint for rapid development of the most powerful and professional military force the world has ever known. Gaining early approval of the Joint Chiefs of Staff and the Secretary of Defense, in a few short years such a force had indeed been developed. General LeMay's sense of priorities, his unwavering determination that they not be comprised, and his personal involvement and leadership made it all possible.

General LeMay's efforts to understand the enemy completely, in terms of capabilities and intentions, provided the basis for continually increasing the size of the forces to meet the increasing threat, as well as developing the force characteristics which assured its effectiveness. During his leadership SAC continually increased its margin of superiority over the potential adversary. There was never a doubt that his Strategic Air Command understood its mission, and was thoroughly equipped and trained to perform successfully.

I flew with General LeMay on many occasions and was always impressed with his superb ability as an aviator, as both pilot and navigator. In all the many occasions I served as his copilot, the only time I was permitted to take off and land the aircraft was when he was taken ill on a trip.

I'll never forget, when we were first considering air refueling using the old B-29 hose system, General LeMay asked me what I thought of air refueling as a tactic for range extension. In my infinite wisdom I advised that, while it had application to a very few unusual missions, we should never depend upon it as a method of range extension for total force use. Years later, when air refueling had become commonplace with both bombers and fighters, he reminded me that my foresight had been slightly flawed.

Lt. Gen. James V. Edmundson

More than any other one man, General LeMay won the war in Europe. The Strategic Bombing Survey, as well as the testimonials of the surviving German leadership, tell us what a vital part of the entire operation the air war was. General LeMay *made* the air war in Europe. As a colonel he was commanding an Air Division — a two-star job. He developed the bombardment tactics, he developed the strategy, he helped pick the targets, and he built the organization.

He was transferred from Europe to the China-Burma-India theater. He took over the B-29 program, putting it on its feet and introducing many of his B-17 concepts into the program. In the Pacific he introduced such innovations as incendiary bombing, low level missions, lead crews, and more. He had won the air war against Japan before the A-bomb ever was dropped.

After being instrumental in winning the air war in three theaters, he was assigned to the Air Staff to head Research and Development. There he had a primary role in initially bringing the strategic missile program from ideas to drawings. He then went to Europe, where he put the Berlin Airlift into motion. This was the most important definitive action taken by the military within the Cold War framework until that time. He used a form of air power, airlift, to achieve a political victory.

Then he built SAC. No one who is aware of the facts of life can see how we could have survived 25 critical years without SAC. The only reason Khrushchev decided to take his missiles out of Cuba was because he knew General LeMay's SAC was looking down his throat.

Finally General LeMay served as Air Force Chief of Staff. He was thwarted in this job, which he was so superbly qualified to do, by President Kennedy and his

New Frontiersmen, and Mr. McNamara and his Whiz Kids. Never had such a grossly unprepared gang of youngsters been thrown into the arena to face problems that were much too big for them. America is still paying the price for having selected a bunch of untried kids and their hangers-on to do a man's job. It would have engendered intense bitterness in a lesser man than LeMay.

General LeMay's absolute dictum that commanders fly needs to be revealed. This is most important because today's cost-effective (?) Air Force has determined that the cost of fuel and the flying hours in tactical aircraft for combat crews is too expensive to accept. General officers, with very few exceptions, are not permitted to fly. General LeMay built SAC on the premise that the surest way to insure competent SAC Wings was to provide competent leadership, and that leaders who were not intimately familiar with their tactical aircraft could not be competent leaders. He has lived by this rule all his life, and he insisted that his wing commanders do so too.

One Saturday morning when the Boss had most of his senior staff in his office for an informal session, he asked all of them to list the SAC wing commanders from one to ten in the order of what they considered to be their competence. At the same time he sent me to get records from Operations, the flying records of all SAC wing commanders, and the flying time of each commander in the combat aircraft with which he was equipped, over the past six months.

He ran it as a blackboard exercise. The opinions of his senior staff were remarkably similar. With minor exceptions they agreed very closely. Then I was asked to post the top 10 commanders in order of their flying time, and the correlation was almost exact. The Boss asked, "Have I made my point?"

Lt. Gen. Bertram C. Harrison

I first met Brigadier General LeMay in 1944 at Mitchell Field. I was taking a B-24 bomb group to North Africa and was undergoing the prior-to-overseas-movement inspection. LeMay was there only for a short time and had decided to brief officers on bombing as conducted by his units in the Eighth Air Force in England.

We had been teaching that the leader should zigzag from the Initial Point to target and not straighten out for the bomb run until 30 seconds from the target. The idea, of course, was to make it harder for the ack-ack to track and hit us — or so we thought. Very highly skilled and trained bombardiers could perform this satisfactorily. Others could not. It was difficult to maintain close formation at high altitudes. Planes spread out had a large bomb pattern and offered poor protection against fighters. Highly trained pilots could perform O.K.

LeMay explained that time over target *area* was important, that a straight line was the shortest distance between two points, and that straight flight required less time to get to the target. Pinpoint bombing accuracy was everything, and a miss was as good as a mile. He required his crews to fly unwaveringly from IP to target and he had fewer losses and better accuracy as a result.

He was right. Months of training and doctrine for lead crews were thrown out, but we later performed better than other units. He was always so damn right. His reasoning was so correct and so obvious. Why were we required to train so stupidly?

Later I was under LeMay when he had USAFE. Col. Curly Edwinson and I were the commanders of the only combat units in Europe then. I had a C-47 troop carrier wing. Curly had the 86th Fighter Wing (P-51s).

There are many stories I could tell you about LeMay then. The command was small. We operated directly under him. For instance, when my air unit was sent to Cyprus to establish a staging facility to care for Christians who might have to be evacuated from the African continent because of Arab-Jewish fighting, we had no communications with USAFE. LeMay solved the problem. He was a ham radio operator. I had a sergeant who was one also. We took the sergeant to Cyprus and he reported daily to LeMay. "Curt" was LeMay's call sign. The sergeant was "Blocker."

Before the Berlin Airlift really started the Russians had been cutting the various communications links to Berlin for periods to see how disruptive they could be. LeMay called Edwinson and me to Wiesbaden, and we planned an air strike against the Russian Air Force in Eastern Europe. The 86th Fighter Wing was to shoot up and frag bomb the planes on the ground on distant bases, while my unit kicked frag bombs out of the side of Gooney Birds over closer bases. The attack would take place at dawn. The penetration was to be at night at low level. He had submitted his plan to Washington, but they would not approve it.

I don't know how effective we would have been at sneaking under their radar cover at night in 1947, but I am sure we would have caused some panic and a lot of damage. Remember we were the only ones with nuclear weapons then. Maybe we could have settled the whole damn mess then.

LeMay left Wiesbaden to command SAC. He requested me when my tour of duty expired. I had to go through the entire programmed processing. There were indoctrination visits to units, schools, exams, and then finally assignment as an Air Base Commander. I later became a combat group commanding officer, deputy wing CO, wing CO, and division commander.

The road wasn't always smooth and easy. General Power wanted to require me to pay for some unauthorized improvements I had made on the operations offices at Rapid City. LeMay squashed the threat and then established a course for all commanders concerning the use, misuse, authority, and problems of construction funds. Everyone was required to max the course. I think that was typical of the way he operated. His officers were better trained and indoctrinated. If you did the wrong thing after showing you knew better, you paid whatever price there was.

During a commanders' conference in Puerto Rico, I took LeMay snorkling above an old shipwreck off the island of Mona, east of Puerto Rico. LeMay wasn't the greatest swimmer and he had never snorkled. The water was rough and infested with sharks and barracudas. Though he swallowed a lot of saltwater, he dived with us and explored the undersea fantasy. He was uneasy when the sharks appeared. My aide, now four-star General Lawson, lost his front bridge while blowing at a nearby shark to frighten him away. I was anxious about LeMay's safety, but deep down believed no shark would dare attack him.

When I was commander at Loring AFB I was asked to accompany LeMay and others to Eniwetok to observe a test of a hydrogen bomb. I emplaned at Offutt, and LeMay flew the modified KC-97. We landed at March AFB, where he addressed the Non-Commissioned Officers' Academy. We took off immediately thereafter and flew to Hickam, refueled, and went on to Eniwetok. We arrived about 0430 their time so tired we could hardly see. We went to breakfast and then toured the island, the test installation, and all the recording facilities. At about 1800 we were told to be ready to rise and shine at 0330. Then LeMay was asked if he would like to shoot skeet. He shot 99 out of 100. I could not have seen the birds. He had flown the entire flight while the rest of us could snooze or relax.

There are hundreds of things that come to mind when I think of the years I served under LeMay. A few were earthshaking. Some were amusing. Many were awesome. He was firm, audacious, enterprising, and vigorous. I never admired anyone as much and would have gone to Hell and back for him. He is a great man who is an enigma to the general populace.

[Author's note: I know of no one who knew General LeMay intimately who did not feel the same way. I certainly do.]

Maj. Gen. John Kline

My first personal contact with General LeMay was in 1946 when he commanded the United States Air Forces in Europe as a lieutenant general. I was a major, chief of Flying Safety, USAFE. There was a B-17 accident in the mountains south of Lyon, France, wherein the pilot had carelessly flown into the very top of a mountain while on a search mission for another lost aircraft. There was one survivor (tailgunner position-observer). I went to the accident site, investigated, prepared my notes for a report, and returned to Hq USAFE, Wiesbaden. There was an immediate request for me to report to General LeMay's office and relate verbally my findings. Those present were General LeMay, Major General Kissner, Brig. Gen. Jack McBlain, plus others. General LeMay invited me to sit down at the table and "report." He asked me many questions about this accident and other accidents in the recent past — he had only recently arrived at USAFE and was getting caught up on his safety situation, which he had concluded was a real problem in that theater.

He agreed with my findings and even asked me for recommendations. After the meeting, I remember, I was surprised how much at ease I was during the discussion. He had created the feeling that he wanted to know my views and recommendations. He made me feel as if I was important to his staff, that my work meant something. I was very much impressed and immediately gained great respect for him, because "going in" I was afraid and "coming out" I had confidence. He wasn't anything like what his reputation purported him to be.

A month later he called a commanders' conference and I was scheduled for five minutes to review the accident record of the past six months, mid to late 1946. The record was deplorable; there was pilot error and obvious lack of command attention. General LeMay introduced me as "Major Kline, who will tell you what I expect in the way of aircraft safety emphasis." He backed up everything I said, which further endeared him to me as a commander. I couldn't believe it. One simple new policy he instituted was the "main cause": any commander who had an accident would report to him personally within a certain number of days and explain the accident in detail. He was tough on the commanders who had accidents thereafter.

One day I was called to his office. By this time I was working at instituting air traffic control in the European area. In the staff there had been some talk about an airlift from Wiesbaden Airport to Berlin. In the staff two of us, Major Willeford and I, had been working under Col. Jack White on planning a multi-aircraft (C-47) cargo lift. One group of C-47s from the Munich area under Colonel Harrison was deployed to Wiesbaden and started flying the route, day by day, and then including the weekend.

When I got to General LeMay's office the subject was the C-47 and its capability. I had helped run a glider pilot training program in the States (1943-1944) employing the C-47 in a close-in, high frequency landing and takeoff operation. General LeMay asked me, "What is the limiting factor in operating C-47s on an airlift (such as we had started to Berlin)?"

My reply was, "The time to load and unload the aircraft." This started a big argument with some of the other staff members in the meeting. My view was that loading and unloading capacity would be limiting before such considerations as number of aircraft at takeoff, or in the landing pattern, or in the enroute pattern, even during bad weather.

General LeMay passed out instructions to concentrate on loading and unloading, brought in another C-47 group, asked for more aircraft, and with his determination built the airlift and flew it for many months (managed by Jack White and his two majors), until a decision was made to bring in a complete headquarters to manage it (General Tunner and MATS). Of course it got much bigger after that, but General LeMay had proved it could be done and that Berlin could be supplied. A lesser man could not have done it.

General LeMay was reassigned to command SAC, and when I was finished with my tour in USAFE I was assigned to the Command and Staff School. Upon completion of AC&SS (December 1949) I was assigned to SAC Hq Special Projects Branch of the operations staff, headed by J. B. Montgomery. I started and managed the Command Post system, and was fully involved with the construction of briefings under Monty for the next several years.

One of our first major briefing efforts was to be given to the Atomic Energy Commission and became known as the AEC briefing. You had a large part in that one.

[During the first briefing to the AEC, in our control room in Building C, Offutt, with the chairman of the AEC present, plus a Senator, and General LeMay's principal staff, Montgomery was briefing and Kline attended the charts. At one point, Strauss, who I believe was chairman, asked General LeMay a question that involved predicting the outcome of an exchange of weapons, or a "fight," and General LeMay replied, "I've known of only one man in history who could predict the outcome of a fight in which he was engaged, and that was Joe Louis." The questions were much more sensible after that.]

During that discussion I was impressed with the clarity and brevity that characterized General LeMay as he explained points of interest or concepts or procedures, compared with the almost clumsy (relatively) manner and wording that Monty and I worked on for days to put in the script.

General LeMay was forever being scheduled to give briefings or talks to various groups. I became the speechwriter and chart manager for a long period with him and Monty. It became such a chore to General LeMay that I believe he began to partially lose interest in reading a prepared script as I turned the charts. One day, when briefing the Air War College class at Maxwell AFB in about 1951 or 1952, as he finished reading a page he would drop the page on the floor next to the lectern. There wasn't an efficient way for him to dispose of the page. As I remember, it was a "hurry-up" affair and the pages were not in a book and the lectern was too small. It was clear that he was in a hurry and impatient. The question period turned out to be brief when he answered a question of a student colonel, "Don't ever end up in my command."

Shortly thereafter, as we were flying to the next briefing, he was in the KC-97 pilot's seat and called for me to bring the script to the cockpit for his review. This time I had the script bound in a notebook, and as he leafed through it I asked permission to make a suggestion. He invited the suggestion.

I said, "General, I've listened to you speak to people in meetings and in private in your office for some years now. Your words are far more interesting and to the point than the words I've written in this script. Further, the audience would far more like to hear your original words as you think of them, than to hear you read what some staff officer wrote in a script." He thought for a moment and agreed to try it. He was a smashing success — no fumbles, straight to the point, and he knew the inference of the charts so well that he could "talk the charts" while I pointed and turned them. In my presence he never read another script.

One time in 1950, or 1951, Margaret Bourke White, the well-known *Life* photographer, got permission to photograph General LeMay. It was decided to do it in the control room (the first control room in Building C, before the addition to Building C, with the newer control room). The room had a world map on the wall at the north end. General LeMay had scheduled ten minutes or so for the effort. He probably thought it was nine minutes too much.

When he arrived, she was already there, and she began talking to him about aimless things and touching his uniform to "place" it properly. After 30 minutes she had not taken one photograph. He was irritated by this time, and the scowl was most impressive. Finally she began taking photographs at a rapid rate, moving him, moving herself. It was obvious after about 25 photographs that she was "fishing" or "groping" for the right place, pose, or position. Came time for lunch and she was asked to continue after lunch. After lunch she had us bring in a ladder and took some photographs from above, moving General LeMay, moving the lights. Finally, after nearly 100 exposures, she said, "Enough," and General LeMay was glad to get out of there.

We had no idea which pose she thought best until *Life* came out on the news stands. It was a simple, straight pose and expression, and the lights had been positioned so that General LeMay's large shadow very noticeably fell across Russia on the world map. The caption read, "LeMay's Shadow Over Russia," or close to that. It was a really impressive photograph and caught the imagination of all who saw it. I decided that M. B. White was a real professional. I don't know what General LeMay thought.

Many times I've been asked, "Who are the greatest

men you have known?" My response has always been easy, but I've limited my thinking or opinion to the time era with which I was familiar. I've said Winston Churchill probably had the greatest impact on the world in my time, but I never met or saw him, so I only know about him what I've read. Of all the people I've met, General LeMay is the greatest man I've ever been associated with. Greatest in what sense? His impact on the total world, the character, drive, and selflessness that guided his actions and decisions, the decisions he made that were far-reaching and for permanence. Take away some of the training, tactical, and strategic decisions that he made, and allow the opposition's will to prevail, and we would probably have a far different world than we have today.

He had the ability to visualize the heights that crews and forces could achieve, if given proper direction and support.

Maj. Gen. Stuart P. Wright

General LeMay was a superb leader because of his honesty, firmness, consistency, thoroughness, and conciseness. Beyond those, one of the qualities I most admired in Curt was his ability to secure the aid of the scientific community for SAC.

An example of his characteristics can best be explained by one personal experience. After SAC was organized, General LeMay put me in charge of communications. This appointment began when Curt called me into his office one day:

Curt: "Stud, what do you think about the communications of this outfit?"
Stud: "I think it's lousy, General."
Curt: "Communication is the heart and soul of any operation, and I want to be able to communicate with every damn station in the world."
Stud: "Well, let's hook 'em up, General."

I began on this assignment by relaying General LeMay's request to Washington. When the task was complete, I sent a memo to Curt. His reply was "Implement." Curt sent notices to every station across the world giving instructions as to the date and the time to contact Hq SAC. This project was successfully accomplished because of the ability of General LeMay to find a problem and work toward its solution to a satisfactory completion.

I close by saying that I would classify General LeMay as one of the finest men I have ever known.

Lt. Gen. Joseph F. Carroll

As one who never had the privilege of serving in SAC, I still have a great admiration and respect for both SAC and General LeMay. I joined the Air Force (on loan initially from the FBI) as a brigadier general late in 1947, right after the passage of the National Security Act of 1947, which for the first time gave the Air Force autonomy (co-equal with the Army and Navy) as a separate entity in the Department of Defense. Stuart Symington was the first Secretary of the Air Force and Carl (Tooey) Spaatz was the first Chief of Staff. Hoyt Vandenberg was the first Vice Chief of Staff.

As I recall, at that time the key command in the new USAF was the Strategic Air Command, because the just-completed world war had convincingly demonstrated the absolute predominance in modern warfare of the strategic application of air power. I believe there were two principal contenders for appointment as commander of this most important command: General Curtis LeMay and General George Kenney. Both were highly experienced, both had sterling combat command records, both were hard taskmasters with great leadership qualifications. At the time I knew both of them by reputation only.

Later, as Director of the Office of Special Investigations and as Deputy Inspector General for Security, I had frequent contact with General LeMay and came to know him, respect him, and admire him as a person and as Commander-in-Chief of the most important command in the U.S. Air Force and in the Department of Defense. From afar I observed General LeMay mold and develop and expand the SAC into the most combat-ready, the most efficient, the most effective, and the most powerful military command in the history of the entire world. He did this by the sheer force of his indomitable personality, by exhibiting those qualities of leadership which inspire subordinates, by demanding and expecting discipline of the highest order, and by having one single criterion for judging a man's accomplishments — superb and successful duty performance.

The next recollection I have of SAC, or at least its mission, was the B-36 controversy. As you will recall, the B-36 was the huge, 10-engine, intercontinental post-World War II strategic bomber, which never fired a shot in combat, but which enhanced U.S. strategic capability and deterred aggression during the height of the Cold War period. At that time the perennial battle was going on between the Army and the Navy and the new Air Force for the greater portion of the overall Defense budget. The Navy, in desperation, it always seemed to me, singled out strategic bombing as a point of attack against the fledgling Air Force, contending as I now recall that strategic bombing was immoral, and its effectiveness exaggerated.

One day in the midst of this interservice controversy, a member of the U.S. House of Representatives, Congressman Van Zandt, stood up in the House and read into the Congressional Record an "anonymous" letter, whose authorship he refused to disclose. This letter not only supported the Navy's position on strategic bombing, but also accused the Secretary of the Air Force, Stuart Symington, and the Chief of Staff of the Air Force, Hoyt Vandenberg, of certain specific despicable and grievous malfeasances in office in awarding the B-36 contract to Convair, of which Floyd Odlum was president at the time. The famous Jackie Cochran was Floyd Odlum's wife.

Naturally a public charge of this nature evoked newspaper headlines and reflected most seriously upon top management in the Air Force. Since Congressman Van Zandt refused to identify the writer of the "anonymous" letter, I was given the task of identifying the author. To make a very long story short, I finally succeeded in doing so. The author was identified as Cedric Worth, a civilian, special assistant to the Assistant Secretary of the Navy for Air.

After some deliberation, it was decided to turn my report over to Congressman Carl Vinson, the chairman of the House Armed Services Committee, which at that very moment was holding hearings on the controversial B-36 subject. I still remember the concluding paragraph of my report; it read:

"This is not an anonymous document. It was prepared by Brig. Gen. Joseph F. Carroll, Director of the Office of Special Investigations, and approved by Stuart Symington, Secretary of the Air Force, and by the Chief of Staff of the Air Force, Hoyt S. Vandenberg."

Cedric Worth was called before the Armed Services Committee, and ultimately he acknowledged authorship of the "anonymous" letter and complete fabrication of its contents.

As you can see, therefore, while I never did serve in SAC I played a not insignificant role in helping SAC get off to a good start.

Also among my recollections:

In the early 1950s before the Soviets developed ICBMs and nuclear weapons, we could depend on the Distant Early Warning (DEW) line for about two hours' warning of imminent bomber attack. It occurred to me as Deputy Inspector General for Security that all an attacking nation need to do to substantially neutralize SAC was to engage in a limited delay-inducing form of sabotage to hold SAC bombers on the ground until the attacking bombers could arrive overhead. Consequently I devised for demonstration purposes a series of delay-inducing tactics, for the purpose of promoting security consciousness within the Air Force.

General LeMay, ever alert to the vital importance to our national security of the mission entrusted to him, asked me to mount such a simulated delay-inducing sabotage attack, simultaneously against a selected number of SAC bases. He even suggested that I set up my command post for the operation in his headquarters at Offutt AFB. He didn't just wait for my report the next morning, but checked in with me frequently during the night to be apprised of developments — such was his degree of interest in his command responsibilities.

Ever alert to eradicating any semblance of vulnerability in his command, General LeMay seized upon the security deficiencies noted in this operation to establish an elite, highly effective internal security system, within his command.

I cite this incident merely to indicate that General LeMay was not only a great wartime commander, but even in peacetime he was one of the few world-wide commanders who recognized that a constantly maintained and effective internal security was an essential ingredient in the combat readiness of his command.

Later, when General LeMay became Vice Chief of Staff of the Air Force under General Thomas D. White, I was assigned in Wiesbaden, Germany, as Deputy Commander of USAFE, hence had little direct contact with him. At that time a decision was made in Hq USAF to turn over all SAC bases in Europe to USAFE; SAC would retain all of its in-place B-47 mission responsibility, but as a tenant on USAFE bases.

Naturally this necessitated negotiations between SAC and USAFE to insure responsiveness to SAC needs and to have allocated to USAFE an adequate number of base support personnel. This latter point created problems in Spain. Tommy Power was CINCSAC at the time and Hank Everest was CINCUSAFE. I was designated by General Everest as the USAFE negotiator, and I began my negotiations with Maj. Gen. Jim Mooney, commander of SAC's Sixteenth Air Force in Madrid. (I was also a major general at the time).

Jim Mooney and I were unable to agree on an equitable division of the manpower personnel involved, nor were Tommy Power and Hank Everest, and I had to go to the Pentagon, where the final decision was to be made by the Vice Chief of Staff. General LeMay, the father of SAC, was then Vice Chief of Staff, and I feared he would be prejudiced in favor of SAC concepts. I was the sole USAFE representative. Hq SAC was represented by a team of officers headed by a major general. After we both presented arguments in favor of our respective positions, General LeMay decided in favor of the USAFE position.

I cite this incident as an indication that LeMay had the ability to rise above parochial considerations, and to adapt intellectually to the burden of broadened responsibility.

Incidentally, I subsequently negotiated with Butch Blanchard, commander of the Seventh Air Division in England, and we quickly came to an amicable and mutually agreeable solution of the problems associated with the turnover of SAC bases in the United Kingdom to USAFE.

Later I had the privilege of working directly for General LeMay when he became Chief of Staff and I was appointed as Inspector General, USAF. I always found him highly supportive of inspection and investigation activities, perhaps because he himself, in his official and personal life, was the soul of efficiency and integrity.

Another event occurred some time in the 1950s, before I was assigned to USAFE and while General LeMay was still CINCSAC. A SAC B-47 was involved in a sensational accident in Texas resulting in several civilian, as well as military, fatalities. My memory is now dim on how I became involved. Perhaps I was appointed as a member of a special board established to investigate the liability ramifications of the accident. At any rate, General LeMay asked me to stop by his office at Offutt AFB on my way back to Washington and give him a report on our findings. I did so, and told him the findings were that the accident was not caused by pilot error nor by maintenance deficiencies, and therefore did not reflect on the capability of the local commander. I'll never forget General LeMay's reaction. He said, "This is the second serious accident in his command. Either he is incompetent or he is unlucky. In either event, I don't want him."

My tenure of office as Inspector General was rather short-lived. It lasted only about one and one-half years. Bob McNamara was Secretary of Defense at the time, and he decided to consolidate Army, Navy, and Air Force intelligence activities in a newly established Defense Intelligence Agency. All three of the services considered it important that their nominee for the directorship of the new agency be selected. I was General LeMay's nominee. I was selected by Mr. McNamara and thus became the first Director of DIA, and remained in that position until my retirement over eight years later. Consequently in the latter years of my service I was associated with General LeMay only in his capacity as a member of the Joint Chiefs of Staff, but my respect and admiration for him grew immeasurably as I observed him accommodate to his enlarged responsibility without compromising in any way his guiding professional and personal principles.

One incident comes to mind from this period. During the Cuban Missile Crisis I briefed the Joint Chiefs daily on the results of the previous day's photographic reconnaissance over the island and recommended for their approval the photographic reconnaissance objectives for the following day. Under procedures established by higher authority, the film canisters were flown directly into Washington, D.C., from an intermediate base of recovery. Some of my people would meet the incoming plane and immediately transport the film to the National Photographic Interpretation Center (NPIC), where it would be developed and interpreted, and photographic enlargements would be made for transmission to my agency, to the CIA, to the White House, and to the State Department. At the same time commanders in the field such as General Cam Sweeney at TAC were tasked with responding immediately to any emergency, even though they had to await receipt of the reconnaissance results from Washington.

One day as I was briefing the Joint Chiefs, General LeMay said to me: "Joe, this is a hell of a way to run a railroad." He was thinking at the time of the field commanders' responsibility. I answered, "General LeMay, I'm inclined to agree with you, but the rules of the game dictate otherwise. Two weeks ago in this very room the Secretary of Defense established this method of procedure and not one voice was raised in protest. I am merely following the procedures established by higher authority."

General LeMay glowered at me for a moment and then said; "Dammit, Joe, you have just passed the buck back where it belongs."

I mention that incident to demonstrate that General LeMay had the greatness of character, even when in a critical mood, to acknowledge relevant facts.

I sum up my service with and under General LeMay with this observation:

He was the very personification of high integrity and efficiency. He insisted upon the highest level of duty performance, even to the point of being accused of unreasonableness by some, as he knew well that so-called "reasonable" standards beget only mediocrity. He had the God-given ability of making people within his jurisdictional influence capable of performing often over and beyond their own innate personal capabilities, simply because those people were part of a superb organization.

Maj. Gen. Robert O. Cork

I don't have much on General LeMay as a cadet. We went to March Field together in 1928. I remember he was a no-nonsense type of guy. I never saw him at any of the swimming parties or beer busts, in which most others participated. Later on, when I was in Panama in the late 1930s he came through as the lead navigator of an experimental flight from Langley AFB, Virginia, to South America, mostly over water. It was like LeMay to make himself the best navigator in the Air Force, probably, and that was before the days of formal navigator training.

At Hq SAC, when I wanted to get papers through Tom Power I was very careful to get both your Operations Analysis and J. B. Montgomery's Operations concurrences. Power respected Operations, and I know he respected you and your people.

I don't think that LeMay's imagination and innovativeness have been given full credit. He was quick to grab anything that had a possibility of contributing something. I remember one time he was driving by the Visiting Officers' Quarters and a young pilot dropped his parachute on the sidewalk and took his baggage inside to check into the VOQ. LeMay stopped his car, backed up, and waited for the officer to come out. Then he gave him hell for his treatment of a parachute. He gave him a good lecture that some day his life might depend on that chute, so he should take good care of it.

You have probably heard the navigator incident. Coming from Hawaii one day the half-way ship which the Navy had positioned was at a certain latitude and longitude. The aircraft was on top. LeMay asked the navigator when they would get to the half-way point. The navigator said, "In two minutes; it will be right under the nose." It was there in two minutes but it wasn't under the nose; it was off the port wing, so LeMay gave his navigator hell and fired him. He said, "We could lose a war with 'navigation' like that."

But he did check with the ship to see if they were on station. They apologized and said they had drifted two miles north. They should have been exactly where the navigator had said the ship would be. LeMay called in the navigator, apologized, and gave him back his job.

He had the reputation for being cold, and not a ladies' man. But he let his hair down one night at a Christmas party at Offutt at the club. My wife said he was the most charming man she had ever met.

You may remember we had a weekly poker game. It included Tom Power, Augie Kissner, Jess Auton, and J. B. Montgomery. The wing commanders usually joined when they were at Hq SAC. I'll never forget the bluffing fights Monty and Tom Power used to get into, after they bluffed everyone else out of the game. Kissner kept close records of individual wins and losses, and General LeMay was always a net winner for the year.

On one occasion just after Jess Auton was killed, his name was mentioned, and LeMay blubbered for two or three minutes.

Incidently, LeMay played close to his chest; he didn't bluff very much. He bluffed so seldom that he was usually successful when he tried it.

[Author's note: I do indeed remember the poker games. Augie Kissner invited me to participate. I have never gambled money in my life. I'm not against it, for those that enjoy it, but I do not enjoy it, so I did not join them. They played LeMay rules, which included a maximum bet and that no hand would start after 11 p.m.]

Lt. Gen. A. J. Russell

I served on General LeMay's staff at SAC from February 1949 until December 1952, first as Deputy Director of Personnel and then, from early June 1950, as Director of Personnel. I joined Hq SAC from the Armed Forces Staff College, and I did not know General LeMay prior to my assignment. As a matter of fact, I knew not one soul at Hq SAC when I signed in and started working for Dick King, then Director of Personnel.

I met LeMay, Power, Montgomery, Kissner, and others on my first night at Offutt in a poker game at General LeMay's quarters. My very first impression of him was that he was a tough, pragmatic, careful gambler who played sober poker to win, and did just that.

I became, and remain, a great admirer of General LeMay. I have never known anyone else with his ability to cut through the nonessential and reduce a problem to its basic elements so that the answer became apparent. In those pre-Korean War days of Secretary of Defense Johnson's "cut out the fat" budgets and resultant lower manpower authorizations, SAC was woefully undermanned for its job. The previous commander of SAC had a restrictive manning policy which, by and large, barred nonrated officers from assignment to and utilization in SAC. When General LeMay took command he found far too many crew members trying to be supply officer, maintenance officer, and so on.

Among General LeMay's early actions, he requested USAF to man SAC to manning table authorization, and he gave internal priority to manning combat crews. We built the atomic-capable units, 509th, 43rd, and others, to authorized strength before we built up a non-nuclear unit. We built by using input — we did very little internal SAC transferring. Once a wing was manned we tried to stabilize the unit. Stability made sense then, and it does today.

It didn't make sense then, nor does it 30 years later, to take a nuclear-capable B-50 aircraft commander and ship him overseas to a non-SAC unit, purely on his overseas return date. Luckily, Lt. Gen. Laurence Kuter was Deputy Chief of Staff, Personnel, USAF, and he understood and approved our actions. In those pre-computer days it took a lot of work both internally in SAC and externally with the rest of the USAF to succeed, but we did the job.

We in Personnel worked closely with the staff in carrying out our responsibilities. Our working relations with you and your staff were excellent, and you and your people were very helpful. Your office was of great assistance to the director of Personnel back in 1951 in establishing a rotation program (EAGLE) for B-29 crews in support of FEAF Bomber Command. I know Operations Analysis was most helpful to the director of Personnel in all phases of the B-47 program. I regret I can't put names to the Operations Analysis members who worked most closely with us, but it was 30 years ago and my files went to the Chief Historian at the time of my retirement. I certainly do remember that your people "enhanced their reputations."

Maj. Gen. Delmar E. Wilson

I first met Lt. Col Curt LeMay at Wendover AFB, Utah, in April 1942. The 306th Bomb Group had been organized at Salt Lake City the previous month and moved to the Bonneville Salt Flats to commence combat training. Col. Chip Overacher was the group commander. LeMay was deputy, and I was group operations officer with shining major leaves. As I recall, LeMay stayed with the 306th for about two months before being reassigned to become group commander of the 305th Bomb Group at Salt Lake City. During the time he was with the 306th, we were quartered in tents while the typical World War II type buildings were being erected. LeMay's desk and mine faced each other. I can remember his sitting for hours, puffing on his pipe, twirling a front lock of hair, and engrossed in deep thought. Hardly a word was spoken.

It later became obvious that he was developing a mental image of the most effective plan for preparing our combat crews to fight a strategic war in Europe. He gave verbal instructions on how our crews should conduct bombing and navigation training. Since higher headquarters, Second Air Force, was in the process of organizing a staff and activating new units, few if any training directives were received. Our training program was strictly a bootstrap operation, self-administered.

Helen LeMay and my wife were living in Salt Lake City during our stay in Wendover. Since I had a car at Wendover, LeMay would ride with me on weekends to Salt Lake City to visit our wives. The stretch between Wendover and Salt Lake was about 140 miles. Many rounds trips were made with LeMay not uttering more than a dozen words. My many contacts in later years convinced me that he hadn't changed. A man of few words. Spoke only when necessary. No idle conversation. Always a good listener.

After LeMay left Wendover, my next contact with him was in England. His 305th Bomb Group followed the 306th Bomb Group by about two months. The first B-17 Bomb Group in England was the 97th, and it had conducted about five very shallow penetrations of the Dutch and French coastlines before the 306th began operations. Since the 97th was the only outfit with so-called "combat" experience, selected crew members were sent to the 306th to indoctrinate the new outfit on how to conduct operations. The greatest emphasis was placed on survival rather than how to destroy the targets. I believe this was a result of the commander's theory that in order to survive it was necessary to fly a weaving course from the initial point (IP) and limit the stabilized bomb run to something like 30 seconds. I should mention that the commander led the 97th on their initial missions and encountered practically no fighter opposition and only a little flak.

Unfortunately, the 306th somewhat adopted the tactics of the 97th, with the result that the initial operations of the first units assigned to VIII Bomber Command destroyed more rutabaga gardens in occupied Europe than military targets. We literally threw bombs all over the countryside because the tactics of going downwind, crosswind, into the sun, and flying a weaving bomb run made it beyond the capability of the bombardiers to accurately synchronize on the target.

LeMay realized that those bombing tactics would not produce the desired results. Consequently he directed his lead crews to fly a steady course from IP to bomb release. And, contrary to some of the operations orders from higher headquarters, LeMay's 305th altered some IP selections that would enable more accurate bomb runs to be made. As a result, the 305th initial bombing results surpassed the other groups by a wide margin.

The LeMay bomb run tactic was universally accepted, and I personally believe that LeMay's 305th Bomb Group demonstrated that strategic precise bombing could be effectively carried out in the European theater. After LeMay was reassigned in June 1943 to activate the 3d Bomb Wing, I assumed command of the 305th Bomb Group until the fall of 1943, at which time I became Director of Operations for VIII Bomber Command, under General Doolittle. For the next 10 months in that position I observed General LeMay's 3d Bomb Wing, but I can't recall any unusual events to report.

My next contact with General LeMay was in Guam in the winter of 1944. His XX Bomber Command had been moved from India and consolidated with XXI Bomber Command in the Marianas. I was Director of Operations, Blanchard was A-3, Kissner was deputy, and LeMay became commander of the Twentieth Air Force, formed by consolidation of the two bomber commands. I frequently flew as copilot for LeMay and accompanied him when he visited the B-29 wings commanded by Rosie O'Donnell on Saipan, Jim Davies on Tinian, Roger Ramey on Tinian, and Tom Power on Guam. I can recall several times riding in the back seat of a jeep driven by the wing commander and listening to LeMay expound on his tactics for getting bombs on the target.

Some wing commanders were more oriented toward survival and keeping losses to the minimum than to placing primary emphasis on bombing results. It didn't take General LeMay long to reorient their thinking. Compared to the European theater, strategic operations against Japan were a piece of cake. The flak and fighter opposition were totally ineffective. The greatest concern of the combat crews was losing an engine or two in the takeoff and initial climb-out, and being forced to ditch or bail out.

While I was assigned to the Twentieth Air Force Hq as Director of Operations, I recall a message dispatched by LeMay to General Norstad in Washington giving his intentions to firebomb Japanese cities from intermediate altitudes. As far as I know, this concept was conceived entirely by General LeMay.

I was assigned as Chief of Strategic Air Requirements in the Pentagon from 1950-1954. Benny Schriever was Chief of Development Plans during most of this period. We had made contacts with Boeing to determine the capability and limitations of the B-47 to conduct low approaches to the target area. We also had received inputs from the AEC Weapons Center for providing low level delivery capability for atomic bombs. As a result of our studies, we decided to put a briefing together and present our views to General LeMay. Schriever, myself, and two or three other staff members went to Hq SAC in either 1952 or 1953. LeMay, Power, J.B. Montgomery, Dick Montgomery, Kissner, and probably 30 or 40 staff were assembled in the conference room of the old headquarters in the bomber assembly building. You were present.

Schriever started our presentation and had no more than introduced the subject of low level approach and bombing, when LeMay stood up, grunted, stuck his cigar in his mouth, and stomped out of the room. Schriever continued a few minutes more before Power and a few other general officers left. As I recall Ariel Nielsen was the only one to ask any questions or appear to be remotely interested. Needless to say, I never had a chance to unfold my briefing charts. We put our tails between our legs and headed for our airplane.

Maj. Gen. Charles W. Schott

We see Curt and Helen occasionally at social functions, and he drops by Rockwell occasionally and I have a chance to talk to him at lunch. He was my first flight leader when, after graduating from training as a flying cadet and getting my wings, I was assigned to the 27th Pursuit Squadron, First Pursuit Group at Selfridge Field, Michigan, in July 1931. He was a first lieutenant in those days and I was a brand new second lieutenant. Strangely, my clearest memory of Curt in those days was sitting in a chair in the squadron ready room facing a corner and studying air navigation. His concentration was amazing, but occasionally he'd put down his book and with a "follow me" signal, walk out to the flight line.

After a short briefing, we'd get into our P-12s, signal the mechanics to pull the props, turn the ignition key, and get started. We'd follow Curt to the downwind end of the field, line up in flight formation, and, at his hand signal, take off. After some formation flying practice, we'd get his signal to break up and practice aerobatics. Then back to the field for a short, and sometimes merciless, debriefing on our performance. Curt would then go back to his studying.

During World War II I didn't see much of Curt, as he was Eighth Air Force and I was in Ninth Air Force. After a postwar tour in the Pentagon and a year at the National War College, I was assigned to Hq Eighth AF and once again was under Curt's command. He was then CINCSAC.

I was in his office at Offutt one day when a young lieutenant from a SAC base was called in to tell about an armament accident. The lad explained clearly what had happened and took full blame for the accident, although obviously some of his airmen were at fault. Curt then asked him if he knew of any reason why he shouldn't be courtmartialed and separated from the service.

The lieutenant, who had been standing in a brace, looked straight at Curt and said, "Not if you think it's justified, Sir." Curt replied, "That's all. Return to duty." I'm not sure which one I admired the most.

Maj. Gen. Nils O. Ohman

I was pretty low down on the totem pole in the SAC pecking order, as a division commander — never on the SAC staff. But I do recall some incidents.

As Commander, Third Air Division, I was summoned to Omaha from Guam in 1956. Curt asked, "How was your trip, Swede?" I replied that my three-engined B-29 did all right crossing the Pacific, with the wind! After I explained the point (#2 engine was a real unfixable lemon), Curt called Archie Old and said, "Give Swede a C-54." With "But, Sir" choked off with a wave of the cigar, Archie went off to arrange the transfer of a much coveted C-54 to the Commander, Third Air Division. I felt a momentary love for the man!

On another occasion I asked Curt whom he favored at the Army-Navy football game, and he growled, "I hope they both lose!"

Maj. Gen. William P. Fisher

My first direct contact with General LeMay began with the deployment of the 509th Bomb Group to Tinian and the need to insure it would be supported and prepared to function when the time came. I was assigned to Twentieth Air Force Hq in the Pentagon and made direct liaison officer between Brig. Gen. Lauris Norstad and Lt. Gen. Leslie Groves (the Twentieth AF and Manhattan District Hq), with the job of coordinating the transfer of the 509th to Tinian in the Marianas. At Tinian it came under LeMay's direct command for the purpose of delivering atomic bombs to Japanese targets when the bombs were ready.

Up until this time the 509th had been trained and equipped almost completely under the direction of the Manhattan District, General Groves, who had his own ideas about how it was to operate. Since it was essential to put the group in operation against Japan before it actually received atomic weapons, to insure that the crews could deliver them accurately, a cover plan was set up. This included delivering TNT bombs (having the same weight, size, shape, and ballistics as atomic bombs) to Japanese targets, and using the same tactics and procedures needed for the real bombs. These bombs weighed about 10,000 pounds each, and LeMay started using them as "block busters" to train crews to deliver them on similar targets chosen for the first atomic bombs. Since the whole operation carried the highest security, only LeMay and a few of his staff were briefed by me on the real mission of the 509th, the projected targets, and the timing of the availability of the first "A" bombs.

There was some skepticism as to whether the bomb would work, as it was really beyond belief at that time. After the Alamogordo test, however, intensive training missions were carried out using the TNT bombs, and after several failures the crews began to hit targets successfully. Tactics were refined by LeMay and the 509th until they were able to routinely penetrate Japanese defenses at operational altitudes by individual aircraft. Since the production of "A" bombs was only about two per month, it was absolutely vital that they be delivered accurately when they became available.

During all this time I was keeping LeMay and his staff advised on weapon status by visits to Guam and bringing back information on the 509th progress and capability to General Arnold and General Groves. By the time the first bombs were produced and available, the 509th had been thoroughly prepared by LeMay, and everything was ready for Hiroshima and Nagasaki.

While operating the FEAF Bomber Command during the last year of the Korean War, I had a fine Operations Analysis staff who were extremely valuable in analyzing the enemy fighter capabilities and particularly their weaknesses, which enabled us to operate our B-29s with very low losses against the large MIG forces based across the Yalu River in China. The details of the Operations Analysis efforts and accurate analyses as to what caused the results of MIG attempts to engage our B-29s enabled us to confidently continue successful operations until the war ended.

Back to SAC and my tour as Inspector General, SAC: Uncle Sam had developed a camera system carried by free helium balloons, air-recovered by a TAC transport group. It was to be used to photograph the USSR. USAF gave the job to LeMay, as the operation was global in scope and they knew he would give it his best shot. We all (his staff) sweated it out, as nobody wanted to operate such a can of worms. The Old Man told Butch Blanchard and me to make a plan to set it up. One night in the Omaha little theater General LeMay pushed by me to his seat and with a grin said, "You're it!"

Col. Walter Coons

As you remember, when we got the first B-47s, they were stationed at Orlando. Col. Mike McCoy had the wing. As I recall, he lost his first B-47 on landing (not the first he received) because the pilot had shut off the main fuel valve as he was preparing to land. That happened because on the fuel manifold were seven switches all lined up, and the young pilot inadvertently hit the fuel shut-off, which shut off the fuel to all six engines. This problem was later rectified by putting a safety cap over the fuel cut-off switch.

McCoy was called to Hq SAC to explain to General LeMay what had happened. As you can imagine it was a traumatic experience for McCoy. He explained what had happened. General LeMay said, "Mike, I presume that if I gave you a baboon, you'd put him in the front seat."

McCoy thought for a few seconds, and then said, "General, under your policies if he had the right MOS (Military Occupational Specialty) I'd have no choice but to put him in the front seat."

With that, General LeMay's cigar switched a couple of times, and he said, "O.K., Mike. Get the hell out of here."

Stanley J. Lawwill

French Red Deer — Maj. Gen. J. B. Montgomery (Monty) delighted in telling the following story of one of those very rare circumstances in which General LeMay was clearly nonplussed.

Once on a staff visit to Europe General LeMay and Monty were invited to hunt on a privately owned game reserve in France. Their guide was a fastidious little Frenchman who explained to them how quietly they must approach the open areas in the forest, which is where they should expect to find deer. He explained that this preserve was stocked with the Red Deer of France, which is not at all as stupid or as careless as the American Mule Deer or Whitetail; and any noise at all would surely defeat the hunt.

As they tiptoed slowly through the woods, the guide recurringly shushed Monty because his cloth jacket made a slight sound whenever a bush brushed across it (General LeMay and the guide were "professionals" — they were wearing chamois jackets, which were silent in such circumstances). Finally the guide directed Monty to stay discreetly behind so the noise of his jacket would not reduce LeMay's prospects for walking up on game.

As they neared a sunny glade covered with waist-high grass, the guide raised his hand — the signal to stop and be absolutely quiet. Sure enough, there in the grass a deer stood up, ears spread, listening for danger. The guide motioned for LeMay to shoot. He did; the deer went down and immediately stood up again. As Monty tells it, "The Old Man was really upset; he couldn't believe he had missed an instant kill even though it was a long shot." (General LeMay is an excellent marksman and has hunted extensively).

General LeMay shot again before the deer could run. Down it went. When they reached the glade, there in the grass were two dead Red Deer. Knowing full well the honor that had been rendered them in being permitted to hunt on this exclusive estate for the carefully preserved Red Deer, General LeMay tried in vain to explain that he hadn't realized that there could have been two deer lying in the grass. He thought he had made a bad shot the first time and had been resolute he wouldn't miss the second time. Monty says, "You should have seen that little Frenchman nodding exaggeratedly (with clear disbelief showing in his face) all through the Old Man's explanation!"

Gen. Bruce K. Holloway

At one of the annual commanders' conferences held at Ramey AFB, Puerto Rico, in the early 1960s, I had a rather memorable experience playing golf with General LeMay and Secretary Zuckert. It happened on the second day of the conference, and for reasons long forgotten I had to do some work in the afternoon and could not rush down to the first tee with the thundering herd (a usual procedure) as soon as the formalities of the morning were over.

When I arrived at the golf shop about midafternoon, Mr. Zuckert, in full technicolor, was very much in evidence and looking for a game. There was one other person there also looking for a game, so we made a quick but complex betting arrangement and took off. The first six holes or so were quite uneventful, except that it seemed to be a contest of who could flub the most shots. Then, on the seventh hole, a golf cart with two people aboard came roaring over the hill and bore down on our threesome.

The driver, whom I recognized immediately, was the club pro, but the other occupant looked like a disgruntled deputy sheriff who had just lost a chase, with a stub cigar in his mouth and cap pulled down over his forehead. I did not recognize him until they had pulled up alongside and the pro inquired, "Do you guys mind if General LeMay joins you?" Mr. Zuckert made some kind of humorously snide crack about bums on the golf course, and General LeMay immediately got out, put his clubs on my cart, climbed in and said, "Let's go."

So off we went. There was no further conversation in our cart for about three more holes, when, feeling compelled to say something (even though enjoying the quiet), I said, "Didn't know you played golf." Curt replied, "It was too rough to go fishing." There was no more conversation for about four more holes until we had to park our cart on the opposite side of the green from where our balls were. Mine was sightly off the green, and his was on, but a long way from the pin. We both got out, and I took my putter and a short iron around to the other side. After I made my shot, General LeMay asked if he could use my putter since he had left his back in the cart.

Lo and behold, he sank the putt, which was about a 40-footer. Without change of expression, and after all three of us had chided him a little on being such a "sleeper," he handed the club back to me and said, "Pretty good putter." I offered to let him use it the rest of the way around, but he just waved it off.

As best I can remember, there was no more conversation until we left the 18th green, when again I commented on his golf, saying that for someone who rarely played he didn't do too badly. He replied that it had been 11 years since he had played with anything approaching regularity, whereupon I stuck my foot in my mouth with some such corny remark as, "Then you did remarkably well."

General LeMay, along with his many other outstanding attributes, is one of the all-time greats of my experience for getting in the last word. That day was no exception. He scowled a little, shifted the cigar to the other side of his mouth, and said, "Yeah; but I couldn't play any better then either!"

It was a thoroughly enjoyable afternoon.

Maj. Gen. Haywood S. Hansell, Jr.

I have been a strong supporter of Operations Analysis in all my active life. I was particularly immersed in bombing probabilities in the early days of air power development at the Air Corps Tactical School. AWPD-1 and AWPD-42 were later based almost entirely upon bombing probabilities in the assessment of force requirements.

During the war itself someone in Operations Analysis (probably a joint venture of the Eighth Air Force and the Hq Army Air Forces Operations Analysis Divisions) did a magnificent job of combining single shot probability deviations of the center of impact of salvo patterns with probable bomb dispersion within the pattern to give a very satisfactory method of forecasting the number of hits on a target from salvo bombing. The process is described in Air Forces Manual No. 79 issued in April 1945 by Hq U.S. Army Air Forces.

I think the nation and the Western world owe a tremendous debt to SAC. I don't think anyone understands what it means to keep a major striking force on combat alert for a decade. But the effort, painful as it was, probably saved us all from domination or a devastating war. The crews of SAC paid a heavy price in terms of their private lives.

Maj. Gen. J. J. Preston

As wing and division commanders our channel was to the numbered air force commander. Only when special conferences were called did we come to Hq

SAC. Of course, on the occasion of an unfortunate accident our contact with General LeMay was swift and direct.

During the period I commanded the 91st Reconnaissance Wing at Lockbourne AFB, Columbus, Ohio, the unit flew the RB-45 aircraft. It was a good airplane; because of newer designs like the B-47, however, engineering updates and modifications were delayed or neglected. As a result, even with proficient pilots the accident gremlin managed to climb on board many a mission.

In March of one year I reported to General LeMay on four different solemn occasions. As was his nature, he listened intently and asked pertinent, penetrating questions. At the termination of the fourth interview he said: "Joe, the 91st Wing will certainly *not* win any Flying Safety Awards this month!"

Maj. Gen. Daniel E. Hooks

A book about General LeMay and SAC by someone who was there would be very valuable, not only for history's sake but also for the pleasure of people like me who were on the edge of what was happening. And you should emphasize the contributions of your crew of Operations Analysts; no one can do it better. Although I was never assigned to SAC directly, I was well aware of how valuable your contributions were considered by General LeMay and others.

I first heard of Curt LeMay in the spring of 1944, when I was getting checked out in B-17s at old Duncan Field, now a part of Kelly AFB, San Antonio, Texas. My instructor pilots were two first lieutenants who had served a combat tour in B-17s flying out of England. They were still a little indignant about the changes LeMay made when he took over their group. Prior to his arrival they would penetrate France a little way, and when the flak would start they would drop their bombs in the fields and go home. But LeMay told them to drop their bombs on target only, and led the group to do just that. If the lead bombardier was not "on target" on the first approach, LeMay would go around for a second approach. The boys said one time he made them fly straight and level for seven minutes to be sure they got on target, and "the flak was hitting us!"

I was on Guam at the Air Depot, Harmon Field, when General LeMay arrived to take command of the Twentieth Air Force. I knew some of the officers who were in his headquarters, and of course heard many tales from them. One concerned General Power, who was a brigadier general commanding a wing of B-29s then (1945). The way I heard it, when General LeMay wanted to quit battling the jet stream at 30,000 feet and wanted to send the bombers over the Empire at night at under 10,000 feet with no ammunition except for the tail guns, everyone opposed the idea except General Power, who volunteered to lead the first such mission. It was a success, of course, and after that I gather LeMay thought very highly of Power.

Another story, which was mostly rumor but was believed by many of us Air Force people, concerned LeMay and the Navy. Again as I heard it, the Navy controlled all publicity releases in the Pacific, and at one time they were said to have used some of the strike photos of the Twentieth bombing raids and passed them off as the results of bombings by Halsey's Fleet! I understood that General LeMay found out about this, and did not trust the Navy for quite a while.

When General LeMay took command of SAC I was working for Howard Bunker at Kirtland AFB in Albuquerque. We were in the forerunner of what was to become the Special Weapons Center. We had close relations with the 509th Group at Roswell. Just from a morale point of view, I would say that the combat capability of SAC doubled in 24 hours after LeMay became commander. And of course it went on up from there.

We knew LeMay was interested in the atomic bomb and all new developments concerning it. But I know of only one visit he made to Los Alamos with members of our organization. On the other hand General Power came through our place several times a year, and sat down with us to discuss new developments and possibilities. I guess I remember such visits because I had known Power when we both were lieutenants and flying instructors at Randolph Field in the middle 1930s.

Lt. Col. Jack Barter

When I reflect back on my early exposure (April 1952) to the SAC Operations Analysis group — you, Lawwill, Ed Berdahl, Marty Schwartz, Rod Smith, and all the rest — my assessment of the impact of that group is that it was unique. And it was unique for several reasons, not the least of which was the character and competence of the individuals (and their motivation), but also the stature that you and General LeMay awarded the group. Since I was a junior major at the time, and also the only "pilotless aircraft" officer in the headquarters, it was natural that I develop ties with your crew, since the questions involving the operational potential of guided missiles to SAC were primarily scientific.

The questions of potential accuracy of ICBMs, and intercontinental range air breathers such as SNARK and NAVAJO, depended on understanding things like reversing gyro drift rates, bending of ground-based guidance radar signals through the atmosphere, and geodesy, shape of the earth, gravity gradients, and so forth. When the time came when we military types had to argue with the peaceful-minded scientists who wanted no part of a strong U.S. defense, but who had academic credentials to impress the politicians, the cutting edges of our arguments were the talents of the Operations Analysis group who could say, "We are a part of both worlds; we understand your academic argument — and it would be nice to trust the Russians — but here are the operational considerations." The academicians couldn't argue with Marty Schwartz, who had sat 13 miles from the 13-megaton Mike shot, in the back of a B-36; or with Ed Berdahl, who rode a jump seat in a B-47 on one of the first Reflex rotations to North Africa; or with any of the others who had done comparable things.

The strength of the Operations Analysis group in the 1950s and 1960s was due in no small measure to how it was used, and its acceptance by the military staff as being mission-oriented. In those days you and General LeMay had a mission-oriented philosophy, and everyone in the command knew what it was. The operations analysts were a part of it because they rode with us on the overseas jumps, slept in the same stinking VOQs at Wright-Pat, or the tents at Frenchman's Flats, or the overheated VOQ barracks at Bolling when the per diem was $9.00 and it cost you $15.00 just to eat. So they were part of the crew, and they pulled their weight.

For me, one of the most useful support activities the Operations Analysis group performed was during our early ICBM test program (1960) at Vandenberg, when we were trying to design the testing rules so they would be statistically significant and applicable to the whole force. It certainly put SAC in the forefront of testing weapons, and it was adopted by the Department of Defense and came out in directive form for the rest of the USAF and the Navy and Army as well. I have copies of the commendation letters from the Weapon Systems Evaluation Group (WSEG) to prove it. For me, that was military operations research at its best.

Maj. Gen. August Kissner

[General Kissner's story is told here in my words — author.]

In 1949 General LeMay took a letter he had received to then Brig. Gen. August W. Kissner, Chief of Staff, SAC. The letter was from the National Geographic Society inviting LeMay to be a member of the Board of Trustees. He said to Augie, "I don't think I'm interested."

Augie read the letter and said, "Boss, reconsider. This is quite an honor. Call Van [General Hoyt Vandenberg, Chief of Staff, USAF]." LeMay phoned General Van, who urged him to accept. He did, and he has been an active contributor to the National Geographic Board of Trustees since then. He continued his participation after retiring, and several years ago he went to the Mideast with the Committee for Research and Exploration.

Appendix A

Pictorial History

A-1

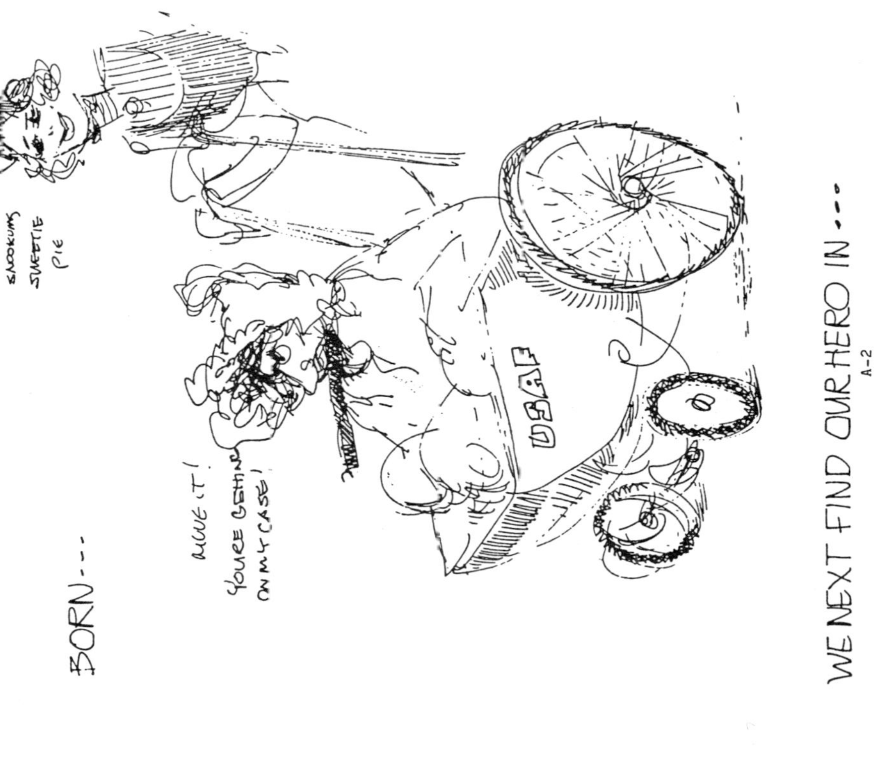
BORN...
SNOOKUMS SWEETIE PIE
LOVE IT! YOU'RE GETTING ON MY CASE!
USAF
WE NEXT FIND OUR HERO IN...

A-2

SCHOOL...
DUKE
MIAMI
MAUL 'EM MIAMI
DESTROY THEM DUKE
BOOGIE AND RUN GEORGE WASHINGTON
SMITH
AT INSTITUTIONS OF HIGHER LEARNING...

A-3

RESIDENT OF LOTUSLAND!
HI!!!!
CLARK! TY! GARY! CARY!
GINGER!
ARTHUR! FRED! GRETA!
HEDDA! LOELLA!
HARRY! LOUIS!
DAVID! BORIS!
JOHN! CHARLIE!
PLEASE! NO AUTOGRAPHS OR SLEEVE TUGGING!
WHERE DO THE FOOTPRINTS GO?

A-5

ACADEMIC
SAY, PROF. WHAT SCHOOL'S THIS?
HE'S SUMMA CUM LAUDE IF EVER I SAW ONE!

A-4

IF THIS IS A RADIATION LAB
HOW COME THE RADIATOR
DOESN'T WORK ???...

A-6

SOLDIER...
BET HE CHEWS BEECHNUT GUM!
HOW DO!

A-7

INVENTOR
Dr. Zimmerman
CHIEF SCIENTIFIC ADVISOR IX TAC
THIS WILLSTOP THE HUN!
SCIENTIST

A-9

ADVISORY LAISON
HIS MAJESTY'S MESS
BLOODY COLONIAL
KEEP IT STRAIGHT
DRESS IT UP!
MOVE IT!

A-8

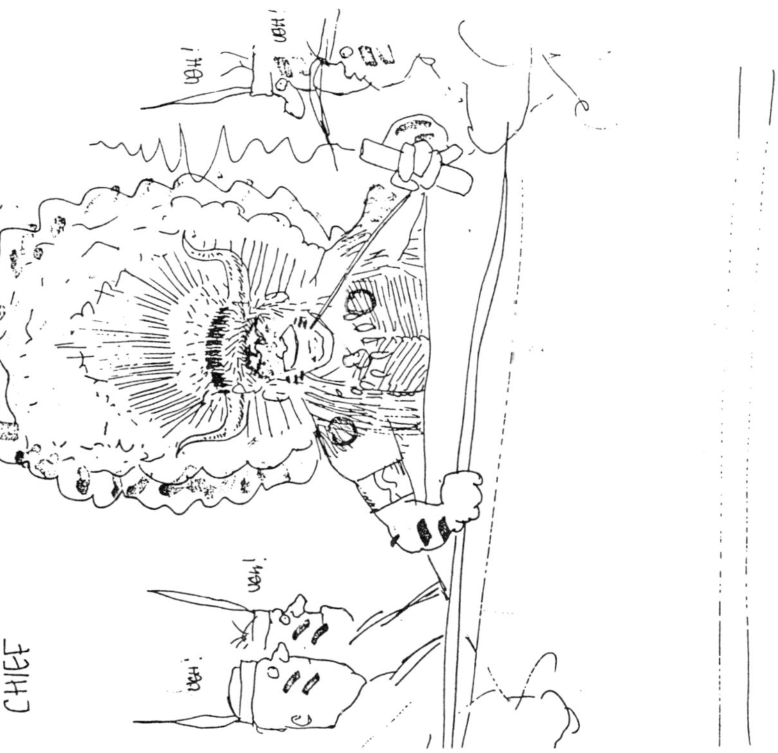
CHIEF
I AM THE CHIEF!
YOU ARE
INDIANS
UGH!
UGH!
UGH!
UGH!

A-11

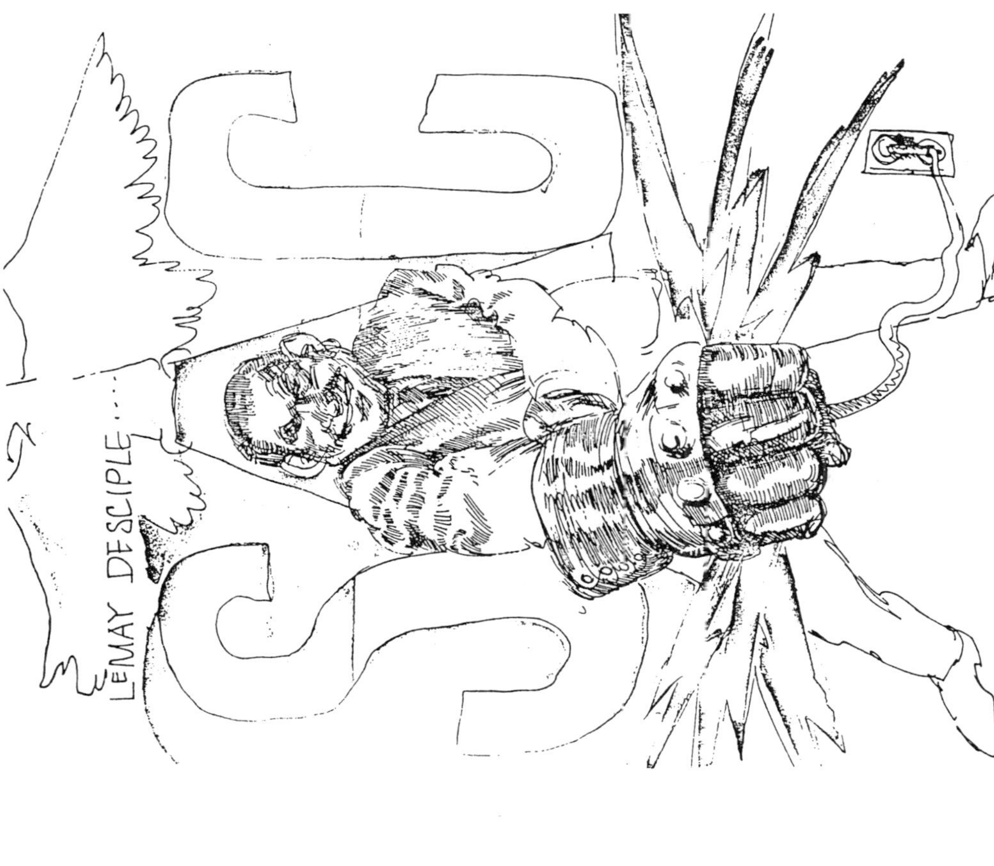
LEMAY DESCIPLE....

A-10

WILDERNESS ADVENTURER
ZIA'S CAMPSITE
A-13

HORTICULTURIST
MOW THAT SWARD!
A-12

HOW ARE SALES FOR VOLUMES I THROUGH XXII OF SABER VOLUNTEER? BY DR. ZIMMERMAN!
AUTHOR
SALE

A-15

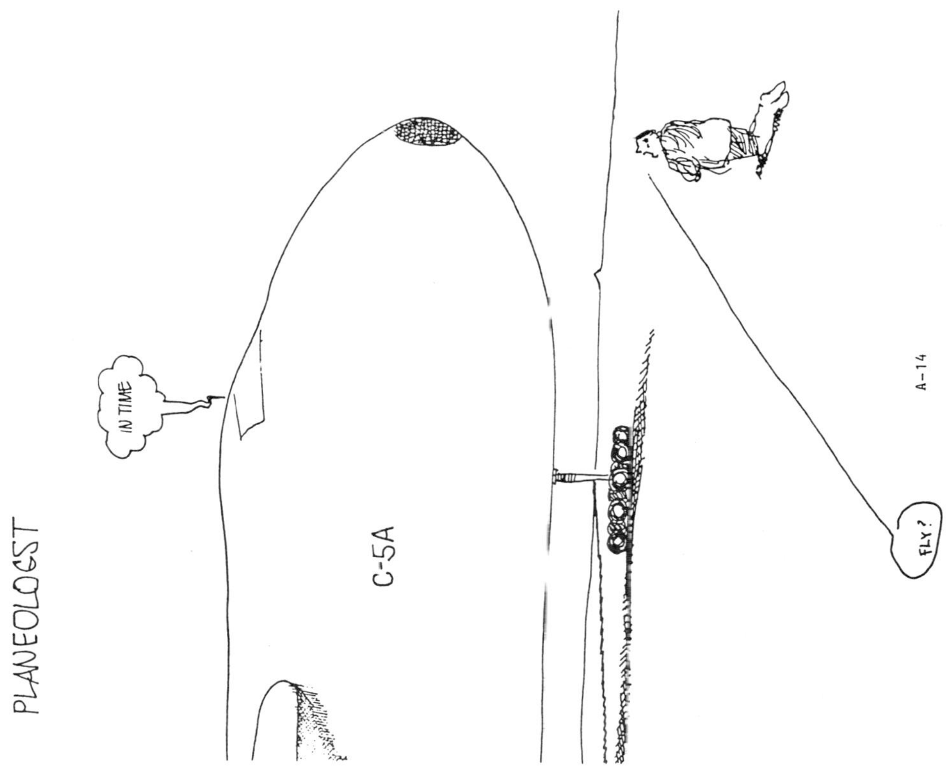
PLANEOLOGST
C-5A
IN TIME
FLY?

A-14

MEIN HERR DOKTOR ZIMMERMAN'S KLINICK UND INZTITUTEZ FUR UNTERZTANDIG UND KOMPAZZIO
VE HAF ANALIZT DER KAZZ UNT VE VEEL DOT CHURE OBZERVAZIONS ARR HOBELESS! DOES YOU ALVAYS VERK IN DER VIFE ZIPED BUILTINK?
ANALYST

A-17

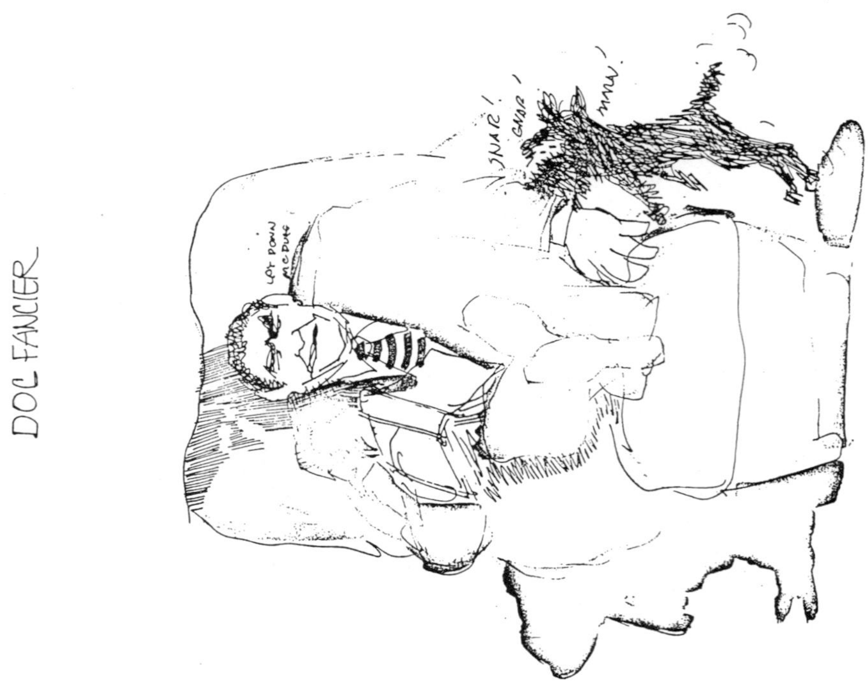
DOG FANCIER
LAY DOWN MCDUFF
SNAR! GNAR!
MMM!

A-16

GOURMET
BEARER! FETCH TEA
YES NABAB TEA WALLAH!

A-18

MOTORIST
I WAS JUST THINKING! I DON'T OWN A MERCEDES, A LAND ROVER, A ZHIGLI, A SKODA, A BUGATTI, A LAMBORGHINI, A CORD, A METRO-BUS, A HILLMAN, A REILLY, A GOLF-CART...

A-19

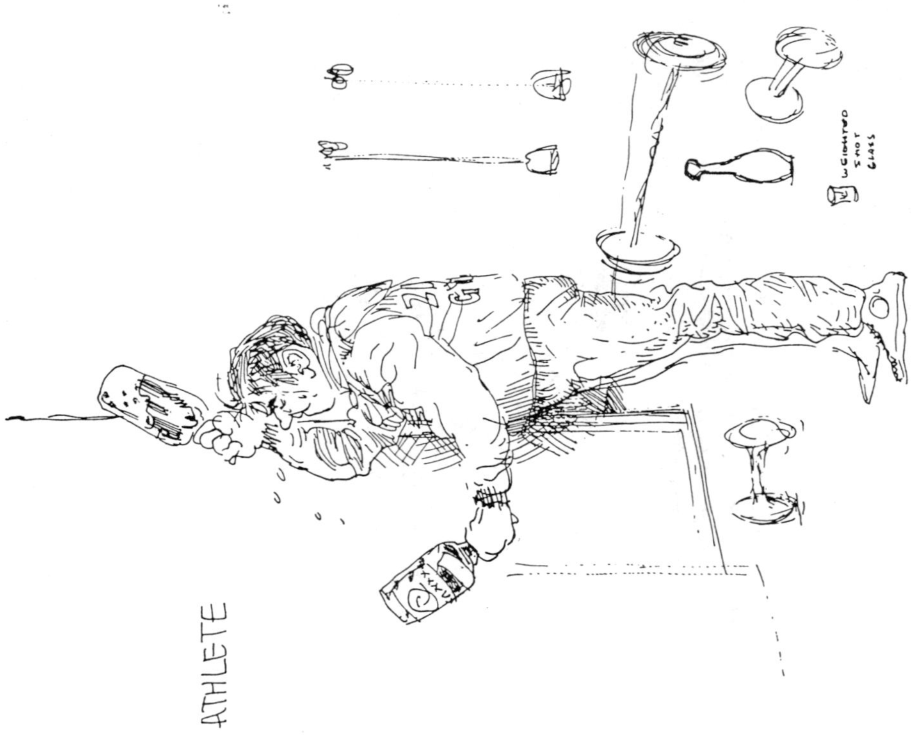
ATHLETE
A-21

POET
MUSE
A-20

Appendix B

Contributors

Many individuals contributed to this book, providing accounts of their experiences with General LeMay and with the Operations Analysis offices within SAC and the USAF, or responding to my requests for information. The book would have been incomplete without their material, and I am most grateful to them. A list of contributors who were not earlier acknowledged follows, alphabetically arranged (the status of each officer is United States Air Force, Retired):

General Samuel E. Anderson
Mrs. Marie Atkinson (for Lt. Gen. Joseph H. Atkinson)
Lt. Col. Jack Barter
John T. Bohn, SAC Historian
General P. K. Carlton
Lt. Gen. John W. Carpenter II
Lt. Gen. Joseph F. Carroll
Robert D. Castater, Chief, Science and Research Office, SAC
General Jack J. Catton
Mrs. Virgil M. Cloyd (for Lt. Gen. Hewitt T. "Shorty" Wheless)
Col. Walter Coons
Maj. Gen. Robert O. Cork
Lloyd H. Cornette, Jr., Director, Albert F. Simpson Historical Research Center, Department of the Air Force
Maj. Gen. Leo P. Dahl
Lt. Gen. James H. Doolittle
Brig. Gen. Thomas J. DuBose
Lt. Gen. James V. Edmundson
General Richard H. Ellis
Maj. Gen. William P. Fisher
Col. Donald C. Foster
Brig. Gen. William Madison Garland
Lt. Gen. Francis H. "Butch" Griswold
Maj. Gen. Haywood S. "Possum" Hansell
Maj. Gen. Bertram C. Harrison
Ray Hemann
General Bruce K. Holloway
Brig. Gen. E. W. "Brick" Holstrum
Maj. Gen. Daniel E. Hooks
Mrs. Paul A. Hower (for Paul Hower)
Maj. Gen. H. E. "Buzz" Humfeld
Col. Lawrence J. Israel
Maj. Gen. August W. Kissner
Maj. Gen. John W. Kline
Col. Bernie Lay, Jr.
General Curtis E. LeMay
Lt. Gen. Richard M. Montgomery
Lt. Gen. George W. Mundy
Maj. Gen. Ariel Nielsen
Maj. Gen. Nils O. Ohman
Mrs. T. S. Olds (for Brig. Gen. Thayer S. Olds)
Maj. Gen. J. J. Preston
Lt. Gen. Donald L. Putt
Maj. Gen. Osmond J. "Ossie" Ritland
Lt. Gen. A. J. Russell
Brig. Gen. LaVerne G. Saunders
Maj. Gen. Charles W. Schott
Col. Raymond C. Scott (for Maj. Gen. Alfred F. Kalberer)
Maj. Gen. Woodrow P. "Woody" Swancutt
Brig. Gen. Paul W. Tibbets
Maj. Gen. Kingston E. Tibbetts
Col. Reade Tilley
General Horace M. Wade
Lt. Gen. Emery S. Wetzel
Maj. Gen. Delmar E. Wilson
Mrs. S. T. Wray (for Maj. Gen. Stanley T. Wray)
Maj. Gen. Stuart P. "Stud" Wright

Appendix C

War Department Special Orders 22 June 1946, No. 139

War Department Special Orders No. 139 dated 22 June 1946 list many important promotions. They give some insight into some U. S. Army and United States Air Force history. Note that General LeMay was promoted from Permanent Captain to Permanent Brigadier General, when he was a Temporary Major General. (The official War Department orders are reproduced here, although the available copy is partly illegible.)

For immediate delivery to Brig. Gen. Curtis [illegible]
referred to in par. 2 *or prompt return to A. G. O., W. D.*

Symbols: DP—By direction of the President
TDN—Travel directed is necessary in military service
WP—Will proceed to
TPA—Travel by officer or his dependents by privately owned automobile is authorized. DS for officer's travel is authorized, par. 1 *e*, AR 605–180, changes No. 3, 31 Mar 1944
AD—Active duty
TDPFO—Temp duty pending further orders
PCS—Permanent change of station
EDCMR—Effective date of change on morning reports

SPECIAL ORDERS No. 139 | WAR DEPARTMENT, WASHINGTON 25, D. C., 22 June 1946.

1. Announcement is made of the apmt of the following-named officers in the RA with grade and date of rank as indicated pursuant to the provisions of section 4, National Defense Act, as amended:

BRIG GEN TO MAJ GEN RA

ROBERT C. RICHARDSON, JR. O1903 USA (temp Lt Gen) with rank fr 1 Oct 1944.
COURTNEY H. HODGES O2686 USA (temp Gen) with rank fr 2 Oct 1944.
JACOB L. DEVERS O2509 USA (temp Gen) with rank fr 3 Oct 1944.
ROBERT L. EICHELBERGER O2624 USA (temp Lt Gen) with rank fr 4 Oct 1944.
CARL SPAATZ O3706 USA (temp Gen) with rank fr 5 Oct 1944.
THOMAS T. HANDY O4665 USA (temp Gen) with rank fr 6 Oct 1944.
MARK W. CLARK O5309 USA (temp Gen) with rank fr 7 Oct 1944.
IRA C. EAKER O7478 USA (temp Lt Gen) with rank fr 1 Dec 1944.
GEORGE C. KENNEY O8940 USA (temp Gen) with rank fr 1 Feb 1945.
WALTER B. SMITH O10197 USA (temp Lt Gen) with rank fr 1 Aug 1945.
[AG 210.1 OO (22 Jan 46)]

2. Announcement is made of the apmt of the following-named officers in the RA with grade and date of rank as indicated pursuant to the provisions of section 4, National Defense Act, as amended:

TO BE BRIG GEN RA

COL BARTON K. YOUNT O2260 AC (temp Lt Gen) with rank fr 1 Oct 194[illegible].
COL JOHN C. H. LEE O2582 CE (temp Lt Gen) with rank fr 2 Oct 1943.
COL WILLIAM H. SIMPSON O2645 Inf (temp Lt Gen) with rank fr 1 Feb 1944.
COL OSCAR W. GRISWOLD O2883 Inf (temp Lt Gen) with rank fr 1 Mar 1944.
COL ALVAN C. GILLEM, Jr. [illegible] Inf (temp Lt Gen) with rank fr 1 June 1944.
COL LEONARD T. GEROW [illegible] Inf (temp Lt Gen) with rank fr 2 June 1944.
COL WADE H. HAISLIP O2774 Inf (temp Lt Gen) with rank fr 3 June 1944.
COL WALTON H. WALKER [illegible] Inf (temp Lt Gen) with rank fr 4 June 1944.

COL WILHELM D. STYER O4415 CE (temp Lt Gen) with rank fr 17 Aug 1944.
COL RICHARD K. SUTHERLAND O4623 Inf (temp Lt Gen) with rank fr 18 Aug 1944.
COL MANTON S. EDDY O4635 Inf (temp Maj Gen) with rank fr 1 Jan 1945.
LT COL J. LAWTON COLLINS O5247 Inf (temp Lt Gen) with rank fr 19 June 1945.
LT COL MATTHEW B. RIDGWAY O5261 Inf (temp Lt Gen) with rank fr 1 Nov 1945.
LT COL LEROY LUTES O5113 CAC (temp Lt Gen) with rank fr 1 Dec 1945.
LT COL LUCIAN K. TRUSCOTT, Jr. O7096 Cav (temp Lt Gen) with rank fr 28 Feb 1946.
LT COL JOHN R. HODGE O[illegible] Inf (temp Lt Gen) with rank fr 1 Mar 1946.
LT COL JOHN E. HULL O7377 Inf (temp Lt Gen) with rank fr 2 Mar 1946.
LT COL JOHN K. CANNON O[illegible] AC (temp Lt Gen) with rank fr 3 Mar 1946.
LT COL HAROLD L. GEORGE O1679[illegible] AC (temp Lt Gen) with rank fr 4 Mar 1946.
LT COL LUCIUS D. CLAY O[illegible] CE (temp Lt Gen) with rank fr 5 Mar 1946.
LT COL ALBERT C. WEDEMEYER O12484 Inf (temp Lt Gen) with rank fr 6 Apr 1946.
LT COL HOYT S. VANDENBERG O15355 AC (temp Lt Gen) with rank fr 30 Apr 1946.
MAJ JAMES H. DOOLITTLE O[illegible] AC (temp Lt Gen) with rank fr 1 May 1946 (RA serial number is O28750).

[S. O. 139, 22 June 1946]

657101—46

2

TO BE BRIG GEN RA—continued

CAPT CURTIS E. LEMAY O17917 AC (temp Maj Gen) with rank fr 2 May 1946.
CAPT LAURIS NORSTAD O18138 AC (temp Maj Gen) with rank fr 3 May 1946.
BRIG GEN RAYMOND S. McLAIN O157562 USA (temp Lt Gen) with rank fr 1 May 1946 (RA serial number is O28751).
[AG 210.1 GO (22 Jan 46)]

BY ORDER OF THE SECRETARY OF WAR:

DWIGHT D. EISENHOWER
Chief of Staff

OFFICIAL:
EDWARD F. WITSELL
Major General
The Adjutant General

THE ADJUTANT GENERAL'S OFFICE
OFFICIAL COPY
WAR DEPARTMENT

[S. O. 139, 22 June 1946]

U. S. GOVERNMENT PRINTING OFFICE: 1946

Glossary

AEAF	Allied Expeditionary Air Forces
AEW	Airborne Early Warning
AFB	Air Force Base
ARDC	Air Research and Development Command
ASM	Air to Surface Missile
CAF	Continental Air Forces
CEP	Circular Error Probable
CINCEUR	Commander-in-Chief, Europe
CINCPAC	Commander-in-Chief, Pacific
CINCSAC	Commander-in-Chief, Strategic Air Command
CO	Commanding Officer
CPX	Command Post Exercise
DGZ	Designated Ground Zero
DOD	Department of Defense
ECM	Electronic Countermeasures
ETO	European Theater of Operations
FEAF	Far Eastern Air Force
GCA	Ground Controlled Approach
GI	World War II Enlisted Man (originated from "Government Issue")
HQ	Headquarters
MATS	Military Air Transport Service
MEW	Microwave Early Warning
MIT	Massachusetts Institute of Technology
MOE	Measure of Effectiveness
MORS	Military Operations Research Society
MP	Military Police
MSL	Mean Sea Level
OA	Operations Analysis
POW	Prisoner of War
PACAF	Pacific Air Forces
PX	Post Exchange
R & D	Research and Development
RAF	Royal Air Force
RBS	Radar Bomb Scoring
RET	Retired
SAC	Strategic Air Command
SACEUR	Supreme Allied Commander, Europe
SAM	Surface to Air Missile
TAC	Tactical Air Command
TDY	Temporary Duty
TWX	Teletype
USAF	United States Air Force
USAFE	U. S. Air Force Europe
VOQ	Visiting Officers Quarters
WAF	Women's Air Force
WSEG	Weapon Systems Evaluation Group
ZI	Zone of the Interior

Index

John J. Vander Velde
Special Collections Librarian
Kansas State University Libraries

Indexer's Note: Individuals may be cited more than once in the index, depending on their military rank at point of indexing and on how they are referred to (form of name used) in the text. — J.J.V.

— A —

— B —